THE FARMER'S GUIDE TO THE INTERNET

Contents Copyright © 1997 by Dr. David Freshwater, The University of Kentucky

All rights reserved. This book may not be reproduced or copied in any form, by any means (electronic, photocopying, or otherwise) without the express written consent of TVA Rural Studies, except for "fair use" purposes such as brief excerpts or quotations in reviews or as otherwise allowed by law.

Library of Congress Catalog Number: 97-61805

James, Henry

The Farmer's Guide to the Internet /
Henry James–3rd ed. / Henry James, Kyna Estes © 1997

Includes bibliographical reference and index.

1. Internet (Computer network). 2. Rural Internet Access. 3. Farming–computers.
I. The Farmer's Guide to the Internet II. Kyna Estes

Third Edition 9 8 7 6 5 4 3 2 1

ISBN 0-9649746-3-0

Book Design: Bill Tyler
Cover Design: Bill Tyler and Al Casciato
Editor: Diane Harney
Printed by TVA Publishing Services

Printed in the United States of America
Printed on recycled paper using soy-based inks.

Published by TVA Rural Studies
400 Agricultural Engineering Bldg.
The University of Kentucky
Lexington, KY 40546-0276

For order inquiries and bulk quantities, write to the above address
or call (606) 257-1872 or e-mail tvars@rural.org.

LIMITS OF LIABILITY AND DISCLAIMER OR WARRANTY

The authors and publisher have used their best efforts in preparing this book, but make no warranty, express or implied, with respect to the accuracy or completeness of its contents. The authors and publisher expressly disclaim any implied warranties of merchantability or fitness for any particular purpose and shall in no event be liable for any loss of profit or any other damage, including but not limited to incidental, special, or consequential damages.

TRADEMARKS

All brand names and product names used in the book are trademarks, registered trademarks, trade names or service marks of their respective holders. The mention of a trademarked name is done exclusively for editorial purposes and to the benefit of the trademark owner with no intention of infringement. Neither TVA nor the University of Kentucky is affiliated with any product or company mentioned in this book.

Farm Journal is the registered trademark of Farm Journal, Inc.
Cover photos copyright 1995 Farm Journal and Photodisc.

ABOUT TVA RURAL STUDIES

The *Farmer's Guide to the Internet* was the first publication of TVA Rural Studies. It was chosen for the initial effort because we believed it was a valuable product in itself. But just as importantly, it served as a model for the operating philosophy of TVA Rural Studies as a whole. To the typical "real world" problem facing rural communities, *The Farmer's Guide to the Internet* offers a market-oriented solution that can be immediately applied by interested individuals. The objective of our entire research agenda is precisely that: searching for practical "people-oriented" solutions that are ready to use.

It is not accidental that issues of rural information access helped define this larger agenda. In November 1994, when the Tennessee Valley Authority Board of Directors established a research center to explore issues shaping the future of rural America, a guiding question was whether rural communities would progress or fall farther behind as a result of the intense pressures of accelerating information and communications technologies. Equally important was the question of how TVA could help rural communities and businesses use new technology to improve their condition.

Seeing a direct analogy to TVA's earliest mission, the Board recognized that electronically transmitted information will fuel wealth generation in the years ahead just as electricity did in the 1930's and 40's. Then, as now, the issue was, "Will access to the most basic raw materials of economic advance be equitably distributed, or will we divide into two America's—plugged in urban populations and disconnected rural ones?"

To deal with these challenges, the objective of *The Farmer's Guide to the Internet* goes beyond helping farmers find quick access to the information necessary for them to run their businesses, although that is a significant goal in itself. The larger intention, from TVA Rural Studies' point of view, is to help build a user base in rural communities sufficient to make the rural delivery of Internet service by commercial companies profitable at competitive market prices.

The farm community is an appropriate starting point. It is, by definition, a business conducted in a rural area, and by general agreement, it is a significant component in the structure of rural life. Many of the farm-related sites laid out in the book will clearly be of interest to rural citizens who are not directly engaged in farming.

TVA Rural Studies will bring the same paradigm and values to a variety of other pressing rural issues: work force development and employment policy, the consequences of electricity deregulation, promoting expanded entrepreneurship in rural manufacturing, reconciling natural resource use with environmental stewardship, and improving our understanding of the forces leading to economic growth and decline.

These are the kinds of issues that will shape the future of rural America. And they will shape it for the better if individual rural citizens are armed with practical solutions like *The Farmer's Guide to the Internet* and the other research products that TVA Rural Studies will work to provide in the years ahead.

David Freshwater
Program Manager, TVA Rural Studies
Professor of Agricultural Economics
The University of Kentucky
Lexington, Kentucky

ACKNOWLEDGMENTS

The Farmer's Guide to the Internet has benefited from the work and insights of many people. We could not have started this book or completed it without their help. We would like to thank all of the people at *Farm Journal* who helped with this project, including (in alphabetical order) Al Casciato, Bob Coffman, Sonja Hillgren, Charles Johnson, Jed Lafferty, and Mary Thompson.

Much of the data for this book was collected under the direction of Dr. David Freshwater, Professor of Agricultural Economics at the University of Kentucky and Program Manager of TVA Rural Studies.

Jonathan Barker, a student at the University of Kentucky, helped us put together most of the technical information in the book, especially the Appendices on each operating system. He also shot the pictures for the book. Jonathan read all of the copy and checked it for technical accuracy. Since Jonathan also created our Web site and maintains our Sun and NT servers and other computer systems, we have found his knowledge and advice to be an invaluable asset to the book.

Kevin Button, a student at the University of Kentucky, was instrumental in researching and compiling the Web addresses for Part II. Helping Kevin with his research were Matt Clarke, a recent graduate of UK, and Brent Frazier and Khris Montgomery, both seniors in the Department of Agricultural Economics at UK.

We would also like to thank the thousands of farmers who bought the first edition and who gave us much-needed feedback via e-mail, letters, and conversations at meetings and in seminars all across the country. Their input helped shape this new edition.

Finally, we would like to express deep appreciation to Norm Zigrossi and Cleo Norman of the Tennessee Valley Authority for their continued support of this project.

Henry James and Kyna Estes
Lexington, Kentucky

FOREWORD

Are we there yet? No, but we're making good time!

Rural America is kicking up its own dust on the fast-paced Internet. Even those on the less traveled back roads are now only a click away from anywhere in the world. Farm families are stepping up their computer power at least as fast as the general public, and such things as local access are becoming a fading issue. They are ready to ride.

As the early Internet pioneers have discovered, it's not just the outreach for information that the Internet provides. It is also providing an opportunity for farm families to find new friends, form new peer groups of common interest, and to bond in exciting new ways, regardless of geography. Some farmers are surprised to learn their next-door neighbor is also exploring the Internet, while others have made lasting friendships and valuable business associations half way around the world.

The pell-mell innovations have come quickly. That's one of the reasons this new, revised edition of *The Farmers Guide to the Internet* was re-edited, barely 18 months after the publication of the charter edition of the best-selling book in March, 1996.

New browsers, new speed, new graphics, new content, new services, and an explosion of new agri-business sites on the World Wide Web have sent the Internet traffic speeding ahead at a rate that can't even be accurately charted. The Internet community has increased four-fold in the 18 months between the original edition of *The Farmer's Guide to the Internet* and the totally new one you hold in your hands now. The World Wide Web agricultural addresses in this book have doubled to more than 2,000 from the original printing. And, it's safe to say, this is not the last word!

The revolution in the creation and delivery of information at the click of a mouse has rapidly progressed from just one of Content to also one of Communication, and the industry now stands on the threshold of re-inventing Commerce on the Internet. It promises to forever change the way business is conducted. Farmers increasingly expect the Internet to save them both time and money and to have increased measurable value to their family and their family business.

Approximately 20% of farm families are already actively using the Internet for a variety of reasons, and it may be conservative to think 40% of rural America will be wiring into the Internet in the next 24 months. The Internet, for them, is quickly shifting from novelty to utility. And, their expectations are running high.

Perhaps no segment of agriculture has taken to the Internet faster than the dairy industry. And, it makes sense. They always had to stay home! Now, they can go anywhere, for information, or for social outreach. "Until the Internet, our best friends were our cows," jokes one Pennsylvania dairyman. "Now, I can talk to someone that can talk back! I can't wait to finish milking each night, and to get online."

For some, getting hitched to the Internet isn't all about farming. Crops farmer Paul Harrison, Menomonie, Wisconsin, found his first online attachment was to the Mazda Miata car club. He ended up in California visiting other Miata owners he had met online. He even helped in an online fund-raising drive for a cancer-stricken Miata owner in Florida. He also follows auto racing through the Internet from his remote location in central Wisconsin.

Sometimes the Internet is about more than business. Age is no deterrent, either. Joyce Lanpher, Onslow, Iowa, is in her early 70's and admits she does the Internet driving for her husband, Lyman. But, she adds that she has been working with computers for more than 20 years! So, the Internet was a natural and exciting extension to the bookkeeping and list management she had been doing. "And it helps that our son, Dave, is a trouble-shooter for other computer owners and can get us out of a fix," she grins. She knows how to locate other folks in a jiffy, using the **www.switchboard.com** address to obtain addresses or phone numbers of people she wants to reach. "E-mail is just part of our daily life," she emphasizes.

Lanpher is not alone. E-mail is the most commonly used activity among online participants. E-mail fits their hectic daily schedules. Pat Corcoran, Chillicothe, Ohio, farms with his brothers in a diverse and geographically-scattered farming operation. He says they often stay in touch by e-mail. Why not?

In a nutshell, Internet usage by farm families is driven by their needs, not by the technology. This is the one book you need to get started, or even for experienced early 'Net users to understand the newest developments in the Internet community. Still, there is nothing like hands-on exploring the wide world that is now at your fingertips.

So, grab the wheel. Enjoy the trip! We're getting there.

Bob Coffman
Farm Journal, Inc.

TABLE OF CONTENTS

Acknowledgements	V
Foreword	VI

Part I

1. Introduction	1
2. Getting Started on the Internet	5
3. E-mail	35
4. Mailing Lists	49
5. Newsgroups	55
6. Chat Rooms	67
7. The World Wide Web	73
8. Agricultural Weather	89
9. Agricultural Market Information	95
10. University and Extension Information on the Internet	101
11. Search Engines	105
12. *Farm Journal's* Web Site (Farm Journal Today)	111

Part II

Internet Addresses (URLs)	119
Internet Service Providers	Appendix A
Rural Internet Access	Appendix B
PCs and Macs for the Internet	Appendix C
Internet Software for Windows 3.1	Appendix D
Internet Software for Windows 95	Appendix E
Internet Software for the Macintosh	Appendix F
Internet Software for Windows 98	Appendix G
Index	

THE FARMER'S GUIDE TO THE INTERNET

CHAPTER 1
INTRODUCTION

The first edition of *The Farmer's Guide to the Internet* was published just 18 months ago. At that time, accessing the Internet from rural areas was a huge problem, Internet software was cumbersome to install and to get to run properly, and there weren't that many farmers using the Net. On the other hand, government agencies, universities and ag companies had begun setting up Web pages, so some information was there. Now, a year and a half later, many more rural areas have local access to the Internet, more farmers sign on every day, and the amount of farming information has more than doubled. The number of farmers using the Internet has also grown dramatically. As this book went to press, the USDA estimated that more than one-third of farms used computers and more than 13% had access to the Internet. At first glance, this number may sound small, but USDA considers a person who owns 40 acres of Christmas trees and also has a full-time job off the farm as a *farmer*, so no one really knows how many "farmers" are on the Internet. In any case, recent surveys show that about 16% of U.S. and Canadian citizens have access to the Internet, so if the 13% figure from USDA is accurate, then it's not too far off from the figure for "average" citizens, most of whom live in cities and work in offices (and don't have a problem with rural access). This is a significant finding because in the past, it has usually taken years or even decades for technologies to diffuse from the cities to rural areas. Yet, Internet use among farmers seems to track closely that of other occupations, even though many farmers still face the challenge of accessing the Internet from rural areas and most don't have a computer where they spend most of their time.

The reason for the phenomenal growth in Internet use among farmers is due to its usefulness as a tool for agriculture. Some farmers are using e-mail to communicate with extension agents, family members, and other farmers; others use it to track commodities prices, and get weather information or reports from university extension services. And, of course, just about every ag company now has a Web site where farmers can get product information 24-hours a day. Having access to information the Internet provides is crucial for today's farming operation because *you can't precision farm without precision information.*

HOW TO USE THIS BOOK

This new edition of *The Farmer's Guide to the Internet* is designed for (1) farmers who don't have a computer and have never been on the Internet, (2) farmers who have a computer but don't have local Internet access, and (3) farmers who are already on the Net and just need an updated list of Internet addresses.

Part I (Chapters 1 through 12) covers everything you need to know in order to use the Internet (how to send and receive e-mail, surf the Web, etc.). Part II is a comprehensive list of farming-related Internet addresses arranged by category. (By the way, this list is updated online at **www.rural.org**.) The Appendices (A through G) cover the technical details of buying a computer, arranging for Internet service, and installing and using Internet software (Netscape and Internet Explorer) on Windows or Macintosh computers. So no matter where you are right now, this book's for you.

HOW FARMERS ARE USING THE INTERNET

Although it seems as if every farming operation is different, in speaking to thousands of farmers over the past year, we have noticed that certain operating practices seem to be gaining in popularity. There are a half dozen or so Internet practices that farmers seem to gravitate to in order to make using the Internet easy and to be able to quickly obtain key information important to their farming operation.

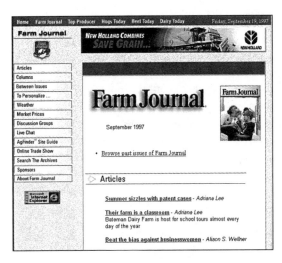

First, farmers find a home page they like and stay with it. The home page is the first place your Web browser (Netscape or Internet Explorer) goes to when it first connects to the Internet. Many farmers point their Web browser to a home page that consistently provides a broad spectrum of useful farming information. A popular home page is *Farm Journal Today*. The address is: **www.farmjournal.com**—see Chapter 7 for information about setting your Web browser's home page to *Farm Journal Today*.

Farm Journal Today is a great site to start from because just about everything you could possibly need is there—news, weather, prices, product information and more. For many farmers, this single address is all they need. (See Chapter 12 for more information about this site.)

The other "best practice" we noticed is that farmers find and bookmark a weather site or two so they can instantly retrieve weather for their farm. Bookmarking is simply a way of saving an Internet address so you don't have to type it in again—see Chapter 7 for more information. The farmers we talked with usually have two main weather sites bookmarked. One is an agricultural weather site (perhaps run by the local land-grant university's college of agriculture.) The other is the weather site from a nearby TV station's home page—see Chapter 8 for more information. By having these two sites, farmers can obtain comprehensive ag-related weather information or simply look at current conditions, forecasts and weather radar.

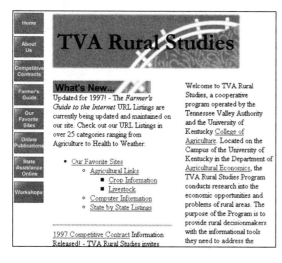

Another common practice was to have a link to a Web site that has lots of links to other farming-related sites. One, of course, is ours (**www.rural.org**), but there are now several other good ones, too. And if you can't find a site you want, you can always search for it using an Internet search engine—see Chapter 11.

The next most frequently used Internet service was e-mail. E-mail is great because you don't have to be at your computer to get it (it will stay in your mailbox until you log on to the Internet)—see Chapter 3.

Many farmers also subscribe to mailing lists or newsgroups that regularly send out information about the type of farming they do (see Chapters 4 and 5), and some occasionally go online to chat "live" with other farmers, extension agents, and ag company representatives (see Chapter 6).

Not only can you obtain and exchange information on the Internet, you can actually conduct real business online. For example, Farm Credit Services' 7th Farm Credit District that serves Arkansas, Illinois, Indiana, Kentucky, Michigan, Minnesota, Missouri, North Dakota, Ohio, Tennessee, and Wisconsin offers an online service called Mainstreet-USA where farmers, ranchers and other eligible rural residents can apply for business loans.

The address is **www.mainstreet-usa.com**. Farm Credit Services' site also provides customers secure access to their account information such as loan balances, tax information and payment amounts. This information is available 24-hours a day, seven days a week, so it does away entirely with the concept of "banker's hours."

You can also access their Financial Services Center for information regarding loans, insurance, leases, and other financial services such as tax planning and estate planning. They even have a "Farm Supply Store" where you can order farm-related merchandise online.

The point is that there is something for everybody on the Internet regardless of the crops you produce or the size of your operation.

CHAPTER 2
GETTING STARTED ON THE INTERNET

WHAT IS THE INTERNET?

If you and I hooked our computers together so that we could exchange information, then we would have, well, two computers hooked together. But if we added a few more people, along with their computers, then we could say we had a computer *network*. Now, let's say that in some other town, another group hooked their computers together; then they would have a network, too. Now (and here's where it gets tricky), let's say that our group and their group connected our *networks* together (using a phone line, for example) so that any of their computers could communicate with any of our computers and vice versa. By doing so, we would have created a network of networks, and that is called an *internet* (lower-case).

By the late 1960s, a lot of companies, universities and government agencies were doing just that—hooking computer networks in one location to computer networks in other locations in order to share information. Of course, these separate internets had their own particular methods of communicating with each other and were not accessible to the public at large (some were even top secret). Over the years, some of these small internets connected with other small internets, making larger and larger internets. One particular internet, called NSFnet, was established in the mid-1980s by the National Science Foundation. NSFnet was used to hook major universities together so that they could share computing resources and information. It became very large, and soon universities were doing more and more of their research using computers, requiring ever more powerful computers and larger networks.

As the power of computers grew, so did their prices. Eventually, major research universities were no longer satisfied with their slower main frame computers and most wanted to buy high-speed supercomputers (which also had a super price tag of several million dollars each). Since the government was paying for most of this research, the government became increasingly concerned with the cost of buying each university its own supercomputer. A young senator from

Tennessee (Al Gore) got the idea that instead of buying expensive supercomputers for all of these universities, the government could save huge sums of money by buying just a few of the costly devices and expanding the NSFnet so that researchers everywhere could have access to the needed computing power over an internet. It was this expanded NSFnet that became what we today call *The Internet* (upper-case). So, there you have it: the Internet is simply a large network of computer networks. And large is the operative term—it is estimated that more than 10 million computers are now interconnected via the Internet, and more than 50 million people (in North America alone) now have access to it. By the way, there are more versions of how the Internet began than there are tales about how Col. Sanders got started. This is our version.

HOW THE INTERNET WORKS:

Now that you know what the Internet is, let's discuss how it works. First, let's assume that there's a computer somewhere out there in cyberspace that has some information you want. This computer is called a *host*. At a party, it's the function of the host to serve you (probably hors d'oeuvres). It's the same thing on the Internet: the host is there to serve you, in this case with the information you request (which is why hosts are sometimes called *servers*). If the "other" computer is the host, then who are you? Well, you are the user or *client*.

It would be a far simpler world if you, the client, could connect directly to the host over the Internet, but you can't. The host is connected to the Internet all the time over special digital lines. Unless you work at a university or a large company that is also always connected to the Internet, you'll have to go through an intermediary to connect to the Internet and get information from the host—usually over a regular dial-up telephone line.

This intermediary is called an *Internet Service Provider* (ISP). Netcom, GTE, MCInet, AT&T, and UUNet are just a few of the better known, national ISPs. Your ISP might also be a local company, such as your phone company or even your electric company (particularly if you live in an area served by telephone or electric cooperatives). Your ISP could also be an online service like America Online, CompuServe, or MSN (Microsoft Network), which provides its own content in addition to what's available to everyone on the Net. For a more complete list of ISPs and online service providers, see Appendix A.

Like hosts, ISPs are always connected to the Internet, usually by special digital lines—lines to which you as a personal or small business probably don't have access. Instead, you'll have to use a line to which you do have access—usually your regular telephone line. There are special high-speed digital phone lines called ISDN that can connect homes and businesses to the Net, and in some areas you may be able to connect via cable TV. However, most people will use their regular phone line to connect to an ISP or online service to reach the Internet.

"MR. SULU, WARP FACTOR 2"

While most people will use their regular phone line to access the Internet, there are sportier ways to cruise the Net. In some areas, special digital lines, called ISDN, are available from the phone company. An ISDN line has a digital channel for voice and another digital channel for data—to give you both voice and data service. These channels can each carry 56 kilobits per second (kbs), and they can be combined to carry 128 kbs of data. An ISDN line costs several times as much as a regular analog phone line and requires a special interface device (regular modems used for standard phone lines won't work with ISDN lines). If you are interested in learning more about ISDN, call your phone company. Another option for higher speeds is to check with your cable TV company. Some of them are now offering high speed Internet access via cable TV. The speed of this service varies, but it can be between 50 and 300 times faster than what a standard telephone line can handle. The costs are usually two or three times higher than the prices charged by most dial-up ISPs, but anyone who has experienced cruising the Internet at 1.5 megabits per second would want it—if they can afford it.

Although the telephone line is the most common way to connect to the Internet from home or most small businesses, the phone system was originally designed to carry the human voice—not the ones and zeros used by computers. For your computer to "talk" over a phone line, its ones and zeros must first be converted into sounds that the phone line can carry. This converter is called a modem. You'll have to have a modem on your end of the phone line and, of course, your ISP will have one as well. So, to paraphrase that old song, your computer is connected to a modem, the modem is connected to the phone line, the phone line is connected to the ISP's modem, the ISP is connected to the Internet, and the Internet is connected to the host. And that's how it all works—more or less. By the way, this description may make the host sound like a mythic or almost god-like computer. Some are truly large and powerful machines, but a host could also be an old 386 PC. The host/client relationship depends on who is serving whom, not on how large one computer is compared with another.

THE MORE THINGS CHANGE...

You've heard the old saying that one year in a dog's life is supposed to be equal to about seven years in a human's life (because people, on average, live about seven times as long as dogs). Likewise, it has been remarked that one "Internet year" is equal to about six weeks of real life because things on the Internet change so quickly. This is a point we tried to keep in mind when writing this book, and it's an important point for you to remember when reading this or any other Internet book. Things change so fast on the Internet that it's very hard to keep up to date.

For example, just as we started writing this book, America Online went from pricing its service by the hour to a flat rate for unlimited access. Flat rate pricing for unlimited access was the future, everyone said. You probably know what happened next: AOL was swamped with more customers than it could handle. To keep up with demand, AOL had to spend an additional $350 million to upgrade its network and was reported to be adding 30,000 modems and phone lines per month.

Also while we were writing, Netcom, a major national Internet Service Provider, did an about-face. It seems that with unlimited access, about three percent of its customers were tying up about one-third of Netcom's capacity. Netcom decided to rethink the concept of unlimited service. One idea under consideration was to disconnect a user's computer if it was online and not being used for some extended period of time. With flat rate pricing, it's not profitable to allow subscribers to tie up a connection for hours on end with no useful purpose. While we can't predict how these problems are going to be fixed, we can expect that ISPs and online services will come up with ways to prevent this.

Some online services used to offer large blocks of time, say 400 hours per month, for a flat rate. While this was more time than most people could use, it didn't sound quite as good as "unlimited" access. I expect that many online services will keep the term "unlimited access" for a flat fee, but will change the definition of what unlimited service means. One way would be to let you have unlimited access as long as you continue to transact business over the Net, but to log you off if your computer sits idle for, say, 10 minutes.

Of course, this is just one idea about how this problem may be solved. By the time you read this, ISPs and online services may have a completely different pricing system. But even then, new issues will arise and new solutions will need to be adopted, meaning things will change again, and again.

The bottom line is that, try as we might, we can't predict exactly how things on the Internet will be working when you read this. For this reason, we are reluctant to recommend particular online services or software packages based on what they currently offer or what their competition doesn't provide at present because these things are subject to very rapid change. On the other hand, the basic ideas behind using the Internet are more stable, and some are even timeless. So don't get hung up too much in the details which change all the time. Instead, get a solid foundation in the basic concepts of going online and searching for and finding the information you need. By having a good grasp of the basic concepts, you'll be better equipped to understand and handle the changes as they occur.

WHAT YOU CAN DO ON THE INTERNET

One problem with trying to describe the Internet is that there are so many things you can do with it, it's hard to know where to begin. One way of thinking about the Internet is to think of it as a *transportation system* for information. In that sense, it's like your family car. You can use a car for lots of things. You can use it to commute back and forth to work, to drive to the grocery store, to take vacations or to take the kids to soccer practice. It's still the same car, no matter how you are using it.

To carry this car analogy a little further, to go somewhere in a car, you need a destination—an address. To go somewhere on the Internet, you'll need an address, too. This address might be the e-mail address of a person or the Web address of a university or company's home page. A Web address is called a URL, which stands for Uniform Resource Locator. The second part of this book covers agricultural Internet addresses or URLs—see page 119.

To finally beat this car analogy to death, if you are in your car and can't find an address or don't know how to get there, you could stop somewhere and ask for directions. The Internet has a similar help feature called a *search engine*. There are several search engines available to help you find addresses on the Internet (see Chapter 11).

Thus, the Internet is somewhat like a car. It's a transportation system (for information) that will take you wherever you want to go (if you know how to get there).

THE FIVE BASIC INTERNET SERVICES

While there are literally dozens of things you can do via the Internet, ranging from video conferencing to playing interactive games with people anywhere in the world, there are five basic Internet services that most people use most of the time. These five services are about all you'll need to access the wide variety of agricultural information available on the Internet.

APPLICATION	DESCRIPTION/USE
E-mail	Electronic mail. Used for communicating directly with other people on the Net.
Mailing List	A service you subscribe to (usually free) in order to automatically receive information (via e-mail) about a particular subject of interest. One popular ag-related mailing list is the Beef-L mailing list. For a list of mailing lists, see page 215.
Newsgroup	An electronic message board where people post messages about a particular subject such as: **news:clari.biz.market.commodities.agricultural,** **news:sci.agriculture.poultry** or **news:alt.agriculture.fruit.** FYI: There are more than 25,000 different newsgroups.
Chat Rooms	Similar to Newsgroups, except that they are <u>live</u>. Once you are "in" a Chat Room, you can use your keyboard to "chat" about a particular subject with other people who are also online at the same time. *Farm Journal* operates several farm-related chat rooms on its Web site. (See Chapter 12 for more information.)
World Wide Web	The most popular service on the Internet. The Web is a graphics-based system for locating and retrieving information using simple "point-and-click" commands (similar to using Windows or the Mac).

As mentioned, there are many other things you can do on the Internet. Some, like FTP (File Transfer Protocol) and Gopher, are older applications that are to a large degree being overtaken by the easier-to-use Web. (In fact, with the newer versions of the Web browsers Netscape and Internet Explorer, you can access FTP sites using the Web browser's simple point-and-click system. (See Chapter 7.) Other services like video conferencing are newer and beyond the scope of this book. If you would like to know more about them, we would recommend Que's *Using the Internet* ISBN 0-7897-0846-9.

THE ONE AND ONLY INTERNET

There is only one Internet. It's not broken down into separate systems for e-mail or the Web, even though people sometimes talk as though it is. If you are using a Web browser to look at a home page, some would say "you're on the Web." But you are really just on the one and only Internet. The services you get from the Internet really depend on which tools (software) you are using and how you are using them.

In the old days of the Internet, all you could do was send and receive text (like e-mail). Then, in 1989, some physicists in Europe created a graphical interface (similar to Windows and the Mac) to help users get around on the Internet more easily. This service looked very different from the plain old Internet, so it was given its own name: the World Wide Web (which is usually shortened to WWW or simply "the Web"). With the development of this graphical user interface, it became possible to use "point and click" techniques to seamlessly move from one area to another on the Web. Because of the graphical interface, the Web could now do more than text; it could also do pictures, sounds, and even video clips and animation.

To navigate on the Web, you use a software application program called a "browser," such as Netscape or Internet Explorer. However, as we'll discuss below, today's Web browsers also handle plain, ordinary text like e-mail in addition to allowing you to browse the Web.

READ ME FIRST!

To connect to the Internet, you'll need six things:

>(1) a computer,
>(2) a modem,
>(3) a regular telephone line,
>(4) Internet communications software (TCP/IP and a dialer),
>(5) Internet application software (a browser), and
>(6) an account with an Internet Service Provider (ISP) or online service like America Online (AOL).

The last item — the decision as to whom your Internet provider will be (either an ISP or an online service) — is extremely important because it affects what Internet software you'll need (if any) and the troubles you'll have to go to in order to get your computer connected to the Internet. In choosing an Internet provider, there are two key questions to ask: (1) do they provide all of the Internet software you'll need (and will they help you install it), and (2) do they have a local access number you can call for free? (If they don't, you may have to make a long distance call each time you want to use the Internet, which can quickly get expensive.)

Most of the national online services like America Online, CompuServe or MSN (Microsoft Network) provide a complete package of software that's easy to install and set up. So do many of the larger ISPs. (See Appendix A for more information.) If you choose to go with an ISP rather than an online service like America Online, we strongly recommend looking for an ISP that will furnish you with a browser package (either Netscape or Internet Explorer) and an installation and setup program that automates most if not all of the tasks involved in configuring your computer and software to work with their computer system. These installation programs are sometimes called Installation or Connection Wizards (or some similar name). Your ISP may furnish this package on a set of floppy discs or on a CD-ROM. In any event, you'll need the version made for your computer's operating system. If you have a PC running Windows 3.1, then you'll need the software for Windows 3.1, not Windows 95 or Windows NT. At the very least, the ISP should provide a manual or other written instructions because the installation procedure requires you to enter information that you can only get from your ISP.

Another reason to choose one ISP over another is the quality of customer support or help they provide. This is, however, very difficult to judge beforehand. Any ISP can claim to offer great support. In most cases, you'll only determine if this is true after you become a customer and need their help.

The range of choices you'll have between ISPs and online services largely depends on where you live. To connect to an ISP or online service, you'll use your regular phone line to call the ISP or online service's access line. If this number isn't a local "free" call, you'll have to pay long distance charges to reach the ISP or online service's access line.

If you live in or near a city, finding an ISP or online service with a local access number won't be a problem. But if you live in a small town or rural area, there may be only a few ISPs or online services to which you can make a free local call. In some rural areas, there may be none. In that case, you will have to make a long distance call to reach them. The cost of the long distance call is in addition to the monthly cost of Internet service. Given that a long distance call can cost 10 cents per minute or more and considering how much time some people spend online, the cost of the long distance call can easily exceed the cost of the Internet service itself. However, there are some ways around the problem of rural Internet access—see Appendix B.

Another consideration is the cost of monthly Internet service. But here you are likely to find much similarity. Most Internet providers these days charge a monthly fee of around $19.95 for "unlimited" Internet access. If you can call their local access number for free (meaning it's a local call), this monthly fee is all you pay.

If the ISP you are planning to use does not provide a complete, easy to install software package, you can use the following information to obtain the software you'll need and install it yourself—but you'll still have to obtain some of the setup information from your ISP.

WHAT YOU'LL NEED TO GET ON THE INTERNET:

A COMPUTER

Obviously, you will need a computer to access the Internet (then again, maybe not—see page 19). If you already own a computer, it may be just fine for the Internet. However, the answer as to whether or not your existing computer will work on the Internet depends on which model of older computer you have—what kind of processor your computer has, how much RAM and hard disk space it has available and which operating system it uses.

YOU DON'T ALREADY OWN A COMPUTER

If you don't own a computer, you'll need to get one. However, you don't need the fastest, most expensive PC or Mac just to access the Internet. It's usually the modem's speed and the phone line's capacity, not the computer's power, which limit how quickly you can send and receive information over the Internet. Even a relatively slow 386SX or 486-based computer can process information faster than it can be sent over regular phone lines. However, if you want to use the Internet to access graphics, pictures, sounds, or video clips, you may want a more powerful, faster Pentium-based PC or Power Macintosh to work with these files. As the Internet gets flashier, you may be happier with one of these newer multimedia-capable PCs or one of the new Power Macs. For a list of basic PC and Macintosh "starter" systems, see Appendix C.

INTERNET COMMUNICATIONS SOFTWARE

Before you can use a browser like Netscape or Internet Explorer to send e-mail or browse the Web, your computer must first be connected to the Internet using the TCP/IP communications protocol. The problem here is that TCP/IP wasn't designed for use over regular telephone lines, and in order to "talk" TCP/IP over a regular phone line, your computer must use a special communications protocol such as SLIP (Serial Line Internet Protocol) or PPP (Point to Point Protocol). You might also run into other versions of these serial communications programs such as CSLIP (which stands for *compressed* SLIP).

As we've said, the national online services like AOL use their own, proprietary methods of accessing the Internet, and the software needed to do that is included on the online service's installation disk (usually a CD-ROM). If you are planning to use an online service (rather than an ISP) as your Internet provider, you won't need to know much if anything about TCP/IP or the SLIP and PPP communications protocols. (The exception is Microsoft's MSN online service which does use TCP/IP and, at this writing, requires Windows 95.)

If you are using an ISP instead of an online service, you'll need to obtain and install a TCP/IP protocol stack and a dialer program (either SLIP or PPP). The TCP/IP protocol stack teaches your computer to communicate over the Internet using TCP/IP while the dialer program does just what it says—it dials the ISP (and "speaks" either SLIP or PPP over a regular phone line). By the way, if you have a choice between SLIP and PPP, the conventional wisdom is to choose PPP. It's said to be slightly faster and more robust.

Many of the larger Internet Service Providers furnish Internet communications and application software as part of a "start-up" package. In most cases, this software has been pre-configured by the ISP and installs and sets itself up automatically with little or no intervention from you. Again, check with the ISP you plan to use and see what software (and setup assistance) they provide before trying to tackle this yourself.

If your ISP can't or won't provide the TCP/IP protocol software and dialer program, you'll have to get it and configure it (set it up) yourself. The specific version of the communications software you'll need to get on the Internet depends on (1) the type of connection (either SLIP or PPP) that your ISP provides, (2) the type of computer you have, and (3) the operating system it uses. Obviously, if your ISP provides a SLIP connection, you'll need the SLIP version of the dialer program, and if they provide a PPP connection, you'll need a PPP dialer. Only your ISP can tell you which protocol they use. Many ISPs support both SLIP and PPP and some dialers support both protocols too.

Newer operating systems such as Windows 95 and the Mac's OS System 7.5 (and up) come with TCP/IP software and a dialer program. You will need to obtain a TCP/IP stack and a SLIP or PPP dialer if you are using Windows 3.1 (or earlier) or an older version of the Mac's operating system. However, certain versions of Netscape and Explorer also come with these programs and these "complete packages" are easier to install and get running. See Appendix D (Windows 3.1), Appendix E (Windows 95), Appendix F (Macintosh), or Appendix G (Windows 98).

In the following examples, we'll show you what it takes to get on the Internet. To simplify matters, we are going to discuss only the most common computers and operating systems and the most widely used Internet communications software. There are obviously other computer systems and other ways to get on and use the Internet, but the information that follows should cover more than 99% of the typical computer systems found in the home or in small offices.

If you are currently running Windows 3.1, pay special attention to which version of Windows 3.1 Internet *application* software you get. Some of the older versions of Netscape and Internet Explorer required that you obtain and install a stand-alone TCP/IP dialer program like Trumpet WinSock. This program was not written by Netscape or Microsoft, and some people have had trouble getting separate pieces from different companies to work together properly. Today, both Netscape and Microsoft make special versions of their browsers that are specially designed for Windows 3.1 and come with their own TCP/IP stack and dialer programs. See Appendix D for more information.

If you have had trouble in the past getting new things to work correctly on your computer, you may want to upgrade your PC to Windows 95/98 or your Mac to OS 7.5 (or whichever version is the latest). Both Windows 95/98 and the Mac's latest operating system have special Internet communications software built in. Alternatively, you may want to use an online service, like America Online, since the software from online services comes with everything you need on one disk and is easy to install and get running.

INTERNET APPLICATION SOFTWARE

Once you are physically connected to the Internet (using the TCP/IP communications software mentioned above), you'll have to have an application program in order to send and receive e-mail, read newsgroups or browse the Web.

In the old days, each of these separate tasks required its own separate application program, but, today, you only need a heavy duty browser, such as Netscape Navigator (version 2.0 or higher) or Microsoft's Internet Explorer (version 3 or higher), any of which can handle everything we discuss in this book. By the way, from this point forward, we are going to call Netscape's browser simply "Netscape" unless we need to specify a particular version such as Netscape Communicator, Netscape Navigator 3.0, and so on. We'll do the same with Microsoft's Internet Explorer, which we'll simply call "Internet Explorer" unless we need to specify a particular version. Most of the time, the particular version (Netscape 3.0, Internet Explorer 2.0.1, etc.) you'll need to concern yourself with will depend upon which operating system you use (i.e. Windows 3.1, Windows 95, Windows 98, etc.)—the rest won't matter.

CONFIGURING AN EXISTING COMPUTER TO ACCESS THE INTERNET

The following descriptions of what your computer must have to run a particular operating system and browser are somewhat conservative. You can, in some cases get by with a less powerful processor, less RAM, etc. Some of the college students we work with could probably figure out a way to browse the Web with an Apple II or a Trash 80 (circa 1978). However, trying to get an under-powered system to work with the latest software can seem like more trouble than it's worth. Finally, keep in mind that things can change rapidly on the Internet. For this reason, please refer to the manuals and other instructions that came with your computer, modem and Internet software. Also, some Internet Service Providers and the online services may have additional information and requirements you'll need to follow to connect successfully.

YOU ALREADY OWN A DOS-BASED PC

While it is *possible* to access the Internet with a DOS-based PC, using the text-based DOS operating system will severely limit the kinds of software you can run and the types of Internet services you can use—you will not be able to view the graphical version of Web pages, for example. Newer Internet browsers like Netscape and Internet Explorer require some version of Windows. At this writing, CompuServe still supports DOS-based PCs, however, this method of accessing the Net can't be much fun. Our advice would be to upgrade to Windows 3.1 if your computer can handle it, or to buy a new PC if yours can't be upgraded. Just about every new PC comes with Windows 95.

UPGRADING TO WINDOWS 3.1

If you have a 386SX PC (or better) with at least 8MB of RAM, you should be able to install Windows 3.1 and still have enough power to run a browser like Netscape or Internet Explorer (or the software from one of the online services). You will also need to have enough free space on your hard disk drive (around 40 MB or so) to hold both Windows 3.1 and a Web browser.

You may have to add RAM or a new hard disk drive in order to run Windows 3.1. While upgrading your existing computer may be an inexpensive solution, some older computers may not be able to physically hold all the RAM you'll need or may have such an under-powered processor that you could easily end up less than impressed with the results of an upgrade. (Some computers may even let you upgrade to a more powerful processor.)

Before deciding on the upgrade route, check the manuals that came with your computer (you did save them, didn't you?) or give the manufacturer or dealer a call to determine what parts of your computer can be upgraded. In deciding whether or not to upgrade, first check out the prices for a new, more powerful PC or Mac. Computer prices have been dropping like stones, and you may be better off in the long run to get a whole new system. (For information about upgrading your old PC or buying a new computer, see Appendix C.)

CONNECTING TO THE INTERNET WITH WINDOWS 3.1

As mentioned previously, the Internet requires that your computer speak in TCP/IP, but the phone system won't let you, so you have to use SLIP or PPP to communicate using TCP/IP. Windows 3.1 was developed before the Internet became widely used. Therefore, Windows 3.1—as it came from the factory—couldn't speak TCP/IP (or SLIP and PPP) and needs an add-on communications program. In days of yore, one of the most popular versions of TCP/IP for

Windows 3.1 was a program called Trumpet WinSock. It used to be that you had to obtain Trumpet WinSock and get it configured properly before being able to connect to the Internet. Some users had trouble getting these earlier versions of TCP/IP communications software correctly configured and operating properly. This situation has now changed. Today, there are special versions of Netscape and Internet Explorer that contain built-in TCP/IP software and a dialer which take the place of Trumpet WinSock. However, make sure you get the correct version!

Since the bulb in our crystal ball went out a few weeks ago, we can't tell you what new versions of Netscape or Internet Explorer will or won't include or what the names and numbers of the new versions will be. You'll need to check to make certain that the version you get says it's for Windows 3.1 and says it includes a TCP/IP protocol stack and a dialer.

At this writing, Netscape users will want either Netscape Navigator 3.0 *Personal Edition* or Netscape Communicator for Windows 3.1. Internet Explorer users will want Internet Explorer 3.0 for Windows 3.1. Older versions of these browsers (such as Internet Explorer 2.1) will also work under Windows 3.1, but require separate TCP/IP and dialer software (like Trumpet WinSock). See Appendix D for detailed information about installing and configuring Netscape and Internet Explorer for Windows 3.1. (Obviously, if you are using Windows 3.1, make sure you also obtain the Windows 3.1 version of the online service's software.)

YOU ALREADY OWN A WINDOWS 95/98-BASED PC

No problem. Windows 95 and Windows 98 come with TCP/IP software. Note that the TCP/IP software for Windows 95 is usually not installed by default (so it may not be installed on your computer, even though it's on the Windows 95 installation disks). To check to see if your computer has TCP/IP software installed and for detailed instructions for configuring Netscape or Internet Explorer for Windows 95, see Appendix E; for Windows 98, see Appendix G.

Again, if you are using an online service like America Online, make certain you get the correct version of their software—in this case, you'll want the version for Windows 95 or Windows 98.

OTHER PC OPERATING SYSTEMS

There are other operating systems such as Windows NT, OS 2, and UNIX that can be used to access the Internet. These operating systems are more often than not used in business rather than for home or personal use and are beyond the scope of this book. If you use one of these operating systems, check with the online service or ISP you plan to use. In all probability, they will have dealt with these operating systems before and may have a solution for you.

YOU ALREADY OWN A MACINTOSH

While you can access the Internet with just about any Mac ever made, it will be easier with newer models and current operating systems than with that old Mac in your basement that you haven't turned on in years. At a minimum, you will need a 68020-based Mac that's running at least System 7 point something and has at least 8MB of RAM (16 MB would be better). You also will need from 4 to 8 MB (or more) of free space available on your hard disk drive, depending on how much disk space your browser needs.

Netscape Navigator claims to need System 7.1 (or higher), while Microsoft's Internet Explorer claims to work with Mac's System 7.0.1 (or higher). However, earlier versions of Netscape and Explorer may work with System 7.0. In any case, as with Windows 3.1, you will also need some kind of TCP/IP communications software in addition to a browser. On older Macs this program is usually called MacTCP (on newer Macs it's called simply TCP/IP) and can be found in the Control Panels. Again, as with Windows, some Macs come with a TCP/IP stack and a dialer already installed (System 7.6 and higher) while some do not. For detailed information about getting your Macintosh ready, willing and able to be connected to the Internet, see Appendix F.

Nearby lightning can create electrical surges that may ride in on the power line (or the phone line) and fry your computer. Plugging your computer and other electronic equipment (modem included) into a surge protector can help prevent damage to your expensive computer equipment. Surge protectors are available at most computer, office supply and home stores.

ACCESSING THE INTERNET WITHOUT A COMPUTER

We said earlier that you'll need a computer to access the Internet. This is true, but there is an exception—sort of. You can access the Internet using a device known as an *information appliance*. One of the first and most popular is from WebTV. While technically it really is a computer, it works (and looks) more like a video game than a PC and it's cheaper and much easier to use than a PC.

WebTV is a box that hooks up to your TV so that you don't need a separate computer monitor. Your phone line plugs directly into the WebTV box (it comes with a built-in modem). And it has its own software (also built in) for surfing the Net. By the way, if you go the WebTV route, be sure to get the keyboard. It comes standard with a TV-type wireless remote which you *can* use to surf the Web, but the keyboard makes the system much more practical.

Set up time for a WebTV system is approximately 30 minutes to an hour. After the equipment is set up, turn the television on and follow the directions in the manual. The direction manual is easy to follow and will explain how to connect to WebTV's servers via your phone line. After connecting to WebTV, you will be asked to type in your credit card number (so the monthly $19.95 access fee can be billed to your card), and then within minutes you are surfing the Web. A WebTV user receives five e-mail accounts and can switch between Internet users. This allows parents to set up an account for children so the children's e-mail goes to a separate mailbox, and the type of material to which the children have access is limited.

In some areas of the country, connecting to WebTV may mean that you have to use a toll or long-distance phone number. This has always been a problem with Internet access costs for rural areas. Fortunately, more and more providers are moving into rural areas so that dialing into the provider is no longer a toll call. WebTV now allows its customers to connect to WebTV using other Internet access providers that support PPP connections and PAP. This allows many people to use a local provider and a local telephone number to dial into WebTV. By using another provider, the monthly charge from WebTV drops from $19.95 to $9.95. You will still be responsible for the monthly charge of the local Internet Service Provider plus the $9.95 charge from WebTV, but you will no longer have the expensive long-distance telephone bills.

Recently, WebTV added print capabilities to its system so that users could print out copies of reports, financial information, Web sites, etc. There still is no way to save reports or information, since the WebTV system does not have hard drive space or floppy drives like computers. While WebTV doesn't do everything a computer can do, it is an easy and cheap way to get on the Internet. Both Phillips/Magnavox and Sony make WebTV boxes which (at this writing) sell for about $300 (with keyboard).

Finally, WebTV updates its own software so you don't have to worry about upgrades. In sum, it's a neat, easy, and low cost way to get on the Internet. The question you should ask yourself is whether you want to put $300 into a box that does essentially "one thing" when, for a little extra, you could buy a starter-level computer that can do much, much more than just the Internet. Sorry, we can't answer that question for you.

For more information about WebTV, call (800) GO WEB TV or (800) 469-3288.

FIVE SERVICES, ONE PROGRAM

Although this book discusses five different Internet services, in practice, only one application program is needed to use all five services—either Netscape Navigator or Internet Explorer (or the similar software programs that online services like America Online or CompuServe provide). These "all-in-one" Internet application programs like Netscape and Internet Explorer are usually called "suites" or sometimes just "browsers" (even though, technically, a browser used to be used only for the Web).

Having a single program handle more than one service is real progress because in the old days, a year or so ago, you usually needed a separate program to access each service. There were e-mail programs for e-mail, newsgroup "readers" for newsgroups, Web "browsers" for the Web, and so on. These separate programs were inexpensive or even free, but setting them up properly and getting them to work together was sometimes a challenge. The latest versions of Netscape and Internet Explorer now do it all, so there's only one application program to set up and only one application program to learn how to use. (This is also true with the software that you use with America Online, CompuServe and most other online services.) Which one you choose (Netscape or Internet Explorer) is up to you—they are virtually identical in features and ease of use. Unless there's a reason for pointing out differences among the various suites like Netscape or Internet Explorer or the online services' similar packages, we are, for the most part, going to refer to these "all-in-one" suites as browsers.

A MODEM

To connect to the Internet, you'll need a modem. First, some background as to what a modem does and why you need one. A computer communicates with other computers using electronic signals consisting of nothing more than ones and zeros. For computers to talk to each other over a system designed to carry the human voice, those ones and zeros must be converted into sounds that can be carried over the phone. And at the other end of the line, those sounds must then be converted back into ones and zeros that a computer can understand. This is what a modem does. It lets computers "talk" to each other over a telephone system made for the human race.

The speed limit on the Information Superhighway isn't measured in miles per hour, but in bits per second (bps). It takes approximately 10 bits to communicate a single character of information—the letter "A" for example—over the phone line. Therefore, at around 10 bits per character, a 9,600 bps modem can handle around 960 characters per second. A typical letter may have 30 lines of text, and each line may have 80 characters, for a total of 2,400 characters. At 9,600 bps, a "typical" letter might take about 2.5 seconds to send (or receive). At 28,800 bps, that same letter could be sent or received in about one second. Not much difference, really.

If all you were sending and getting from the Net was text, then it might not matter if you had a 9,600 bps modem or a 28,800 bps one. However, today, thanks to the World Wide Web, the Internet is graphics intensive. And if a picture is worth a thousand words, most Internet graphics are even larger. A typical home page on the Web may have a dozen graphics, each of which could easily be around 10K in size (that's about 10,000 bytes or about 100,000 single *bits*). Thus, the total size of a typical home page might be 120K or larger—50 times the size of the typical text-only letter cited above. Now the difference in modem speeds becomes important. A 9,600 bps modem will need about a minute and a half to capture this page while a 28,800 bps modem can do it in 41 seconds. Since 28,800 bps modems usually incorporate error correction and compression, the actual download time might only be 20 or 30 seconds. As you click from page to page on the Web, waiting on a slow modem will quickly become annoying. Actually, waiting on a fast modem can be annoying; waiting on a slow one is intolerable.

Given the amount of graphics on the Internet (and newer, even more graphically intensive applications like Java and animation), don't buy a modem that's less than 28,800 bps. A 28,800 bps modem from a top manufacturer like U.S. Robotics costs about $100 (probably less by the time you read this). Modems for regular phone lines run up to 53,000 bps (advertised as 56K), cost less than $200 and usually include free software (like Netscape) and a free trial Internet connection. To go faster than 53,000 bps, you'll need a special phone line called ISDN (which may or may not be available from your local phone company, and can cost several times what a standard residential line costs). Some cable TV companies now offer very high speed Internet access, but again, this may or may not be available where you live, and it usually costs several times what dial-up ISPs charge.

You may already own a modem—perhaps one came with your computer when you bought it. If your computer is more than a few years old, in all likelihood, the modem "included" with your computer is probably not a 28,800 bps modem. Chances are good that it's a slower model (9,600 or 14,400 bps). If your computer is really old, your modem could be even slower.

THE TRUTH ABOUT 56K MODEMS

If you're browsing through your local computer store, you'll probably find a good number of modems that carry a 56K logo. A true 56K connection would be twice as fast as a regular 28.8 modem and would equal the speed of a single channel ISDN line. That's an amazing amount of data for just a single analog phone line! Unfortunately, there are a couple of reasons why 56K modems just can't deliver that kind of speed. First, FCC regulations limit data transmission speeds on a standard telephone line to no more than 53K. Second, the 53K capabilities only work in one direction—for downloads from the host to the client. Uploads, or data sent from your computer to the Internet, can only travel at 33.6K. That's not nearly as powerful as ISDN, but still a vast improvement over 28.8 technology. Of course, when you're at the computer store ready to buy that new 56K modem, what you may not notice is that there are two competing 56K standards. One, called 'X2', is supported by U.S. Robotics, the other, called 'K56flex', is supported by Hayes. Unfortunately for the consumer, these two modem technologies aren't compatible with one another. That means if your ISP supports K56flex from Hayes, and you've just bought the latest X2 modem from U.S. Robotics, the fastest download speed you'll see is 28.8. Before buying a 56K modem, be sure to check with your ISP to find out which brands (if any) they support or recommend.

MODEM FEATURES

Besides speed, there are several other features that can improve a modem's overall performance. Two important features to look for are called V.34 (or V.42) and V.42bis. A modem that supports these two standards can automatically correct errors, run at top speed on noisy phone lines, and compress data during transmission so that you can send and receive files faster. These features don't add much (if anything) to the purchase price of the modem and are available on almost all modems made by the top manufacturers. So, in addition to buying a 28,800 (or 56K) bps modem, get one that supports both V.34 (or V.42) and V.42bis.

INTERNAL VS. EXTERNAL MODEMS

For PCs, you have a choice between either an internal or external modem. Internal modems fit inside the computer case and are powered by the computer itself. An internal modem won't take up any valuable desk space and doesn't have to be plugged into a separate electrical socket. Because it doesn't need a separate case and power supply, an internal modem is usually a few dollars cheaper than an external version.

External modems have their own case and need an external power supply that must be plugged into an electrical socket. An external modem usually has status lights that show how the modem is working—a feature that few internal modems have. External modems use one of your computer's serial communication ports, and in some cases these ports can be slow and can actually rob your modem of some of its performance. On the other hand, installing an external modem is easier than putting one inside the PC.

On the Macintosh side, external modems are the rule, although Mac laptops generally have internal modems.

If you are buying a new modem, look for one of the major brands like U.S. Robotics, Global Village, Supra, or Zoom. Note that modems made for PCs and Macintoshes are not usually interchangeable.

ACTUAL CONNECTION SPEED

Remember the saying that a chain is only as strong as its weakest link? Well, a communications chain is only as fast as its slowest link. If you have a 56K bps modem and your Internet access provider has a slower connection speed (14,400 bps, for example), you'll send and receive information at the slower rate and no faster.

Nevertheless, we recommend buying at least a 28,800 bps modem even if your ISP's connection speed is slower. Sooner or later—probably sooner—your ISP will upgrade to 28,800 bps or faster (or you may change ISPs). In either case, you'll be ready.

A REGULAR TELEPHONE LINE

Obviously, a modem must be connected to a telephone line to access the Internet. But you don't need a special telephone line; a regular home phone line will do just fine. The modem plugs into the phone jack in your house just like a regular telephone. In fact, most modems come with an extra jack so that you can plug a telephone into the modem and still use the telephone. Virtually all modems use the same kind of modular plug-in connector that home-type phones use. It's called an RJ-11, and this is probably the type of phone jack you have in your house. However, if you don't have phone jacks (or have a different kind) you'll need to install a modular type RJ-11 phone jack. They are available from places like Radio Shack and Wal-Mart for a few dollars, and they're easy to install.

A really noisy telephone line can interfere with your Internet connection. If the phone line sounds noisy to you, chances are that the line will "sound" noisy to the modem, too, because the modem uses the same frequencies to communicate over the phone that you do. Have the phone company check your line if it always seems noisy.

TWO LINES FOR THE PRICE OF ONE

While you are on the Internet, neither you (nor anybody else in the family) will be able to make (or receive) phone calls. (If, like me, you use the Internet a lot, this can be an issue with certain teenage family members who shall remain nameless.) To get around this problem, more and more families are getting a second phone line, reserving one line for the Internet's modem and using the other for the family's phone. Depending on your state's regulations, and how much your family uses the phone, you may be able to get a second line for about what you're paying now for just one line! How is this possible? By taking advantage of your phone company's optional calling plans.

The most common calling plan is called Flat Rate service. With this plan, you pay a fixed amount each month, and you can make as many local calls as you want. You're probably on a Flat Rate plan right now; most people are. But many phone companies have an optional calling plan called Message Service. This plan can be much less expensive than the regular Flat Rate plan and might let you get two lines for the price of one (more or less).

With Message Service, you typically pay a monthly fee at about half of the normal Flat Rate charge. You are allowed to make a certain number of "free" calls each month, and after that you pay a small charge for each local call. Look in the White Pages of your phone book to see if the Message Service plan is available in your area. The details vary state by state, but a typical Message Service plan might charge you half the normal Flat Rate fee and let you make up to 30 free calls each month (per line). After that, you might pay 10 cents per call or, in some cases, a few cents per minute.

Most people have Flat Rate service because they don't like the idea of "paying" to make local phone calls. But you may not make as many calls as you think. To prove this point, for one month (or one week and multiply by four) write down each time you make an outgoing local call (incoming calls and long distance usually don't count). Calculate what your total phone bill would be for Message Service, based on the phone company's charges and your family's usage, and compare that with what you're paying now for Flat Rate service. Many people will find that they can get two Message Service lines for about what they are paying now for one Flat Rate line. Note that phone companies usually require that all of your phones be on the same type of service, so you won't be able to have a Flat Rate line and get the second line under the Message Service plan.

CALL WAITING

If you have Call Waiting or a similar feature that "beeps" while you're talking to let you know you have another call, be sure to turn it off while you are online. Call Waiting can interfere with your online session (the modem will think the beeps are data). Most communications programs can be set to turn this feature off automatically by dialing a special code just before dialing the access number.

You can turn off Call Waiting yourself by having the modem dial the code immediately before it calls your online service. The code to disable Call Waiting is usually *70 (the asterisk followed by 70). Your phone company may use a different code (such as 1170). Check the front section of the White Pages of your phone book or call your phone company for instructions about how to disable Call Waiting. To get your modem to deactivate Call Waiting, you have to add the Call Waiting code to the modem's dialing string in the dialer program you are using. See the manual that came with your modem (or dialer program) for more information. Usually, you do not need to enter a special code to turn Call Waiting back on after your modem hangs up; that happens automatically.

INTERNET SERVICE

Okay, you have a computer, a modem and a phone line. The next step is to choose the company that's going to provide you with access to the Internet. There are essentially two types of Internet access providers—online services like America Online and true Internet Service Providers like Netcom or MCInet. What's the difference?

Online services such as America Online, CompuServe or MSN (Microsoft Network) operate their own computer networks and provide their own content in addition to being able to connect you to the Internet. When you first log in to the online service's computer, usually, the first thing you'll see is a "welcome screen." At this point, you are not on the Internet—you are only on the online service's computer network, but you can access any of the information (or content) they provide their subscribers (information that is sometimes not available on the Internet itself). In addition to providing their own content, virtually all online services offer connections to the true Internet.

True Internet Service Providers, like UUNet, AT&T, etc., don't usually provide their own content. Instead, they simply throw you out into cyberspace and it's up to you to know what you are doing and where you are going. On the other hand, you'll get a true TCP/IP connection, and (within reason) an ISP will usually let you use the Web browser you prefer. Some ISPs also provide ready-to-use software that installs easily and many offer connection assistance if you need help. Thus, there is less and less difference between online services and ISPs. In fact, the Microsoft Network (MSN) is really an online service and an ISP.

CHOOSING BETWEEN AN ONLINE SERVICE OR AN ISP

An online service is a good choice if you don't know much about computers, modems, TCP/IP, SLIP, PPP and such. Online services are like the "training wheels" of the Internet. They provide their own software that's easy to install and easy to use. They charge about the same as a regular ISP, and they have menus and other aids to help you find your way around. Since they can connect you to the "real" Internet, you will have access to anything that's there and you can send e-mail to and receive e-mail from anyone on the Net (or on another online service). Thus, an online service is a good choice if you are just starting out and want to get your feet wet before jumping into hyperdrive in cyberspace. Later, after you've gained more experience with computer-based communication systems and the Internet, you could always change to an Internet Service Provider.

America Online's main "Welcome" screen. Although it has been slow or busy because of its fast growth, AOL is America's number one online service because it's so easy to use and has great content.

If online services can connect you to the Internet, then what are Internet Service Providers for? Well, they do the same thing, more or less. While online services can link you to the Internet, the connection is frequently not as fast as the service from a true ISP. This is probably due to the fact that some online services don't provide a true TCP/IP connection. Instead, they use their software to "translate" their system's protocols into TCP/IP—and this takes time. In addition, you typically must use the online service's software to access the Net and this can be a mixed blessing. As mentioned earlier, they give it to you for free, it's easy to install and easy to use. For e-mail and other text-based services, there is really little discernible difference, but for graphics-based applications like the Web, the differences are more pronounced. The Web browser that comes with America Online or CompuServe's software pack isn't as full-featured as, say, Netscape. In theory, some online services claim to let you use other third-party browsers like Netscape with their service, but in practice, this is as (or more) difficult than setting up the same software with a true ISP. Microsoft's MSN is one exception. MSN is an online service that provides a true TCP/IP connection to the Internet and (obviously) you can use a stand-alone Web browser like Internet Explorer. Again, there are trade offs: MSN does not (at this writing) provide the same quality and variety of content as America Online and CompuServe do. This may or may not matter to you. Since just about any Internet provider will give you a month's free trial, you should take advantage of these offers and try before you buy.

E-mail addresses are not portable. If you sign up with AOL, for example, your e-mail address might be "yrname53@aol.com." Later, if you switch to, say, Netcom, your e-mail address would change to something like "yourname@netcom.com." When (as in this example) you cancel your AOL subscription, AOL cancels your e-mail account. Internet providers are under no obligation (yet) to forward your e-mail, so e-mail sent to an old address won't get to you. Changing Internet providers could cause a problem if you've already given your e-mail address to a lot of people. For this reason, it's a good practice to make sure you are going to stay with your current provider before you print your e-mail address on stationery and business cards—or paint it on the side of your truck.

OTHER IMPORTANT CONSIDERATIONS

Another consideration in choosing an Internet provider is whether or not the online service or ISP has a local access line. The local access line is the telephone number your modem dials to connect to the online service or ISP (who then connects you to the Internet). Although you must pay the ISP a monthly fee for Internet service, this cost assumes that you can place a free local call to their access line. Most of the Internet providers listed in Appendix A provide local access lines in virtually every major U.S. city and many smaller ones as well. However, not all Internet providers have local access lines in all areas, so the choice of Internet service may be limited depending on where you live. This can be especially true if you live in a rural area. You may have a choice of "one" ISP or even none that offer local access lines. If that's the case where you live, then you may have to make a long-distance call to connect to an ISP. Given the rates most people pay for long-distance service and the amount of time people normally spend on the Internet, that call will cost you plenty. For rural users, the long-distance charges to connect to an ISP often cost more than the Internet service itself.

If you live in or near a city or town with a population of, say, 50,000 or more, you will probably be able to find one or more Internet providers who have local access lines. It's the people who live in rural areas who have the hardest time connecting to the Net. If you are one of them, see Appendix B for help in getting on the Internet from rural areas.

PUTTING IT ALL TOGETHER

Writing a book about how to use the Internet is akin to writing a book about how to ride a bicycle. Riding a bicycle is not hard, five-year-old kids do it. But imagine trying to write down all you would need to know about riding a bicycle. That could turn something that's very easy (once you've done it) into something that sounds very difficult (if you've never done it).

Writing about the Internet is very similar. By writing down all of the things that are going on and all of the things that could possibly happen, we've made the Internet look complicated and difficult to use. It really isn't. Sure, it's quirky and won't always work as expected. And, yes, you will occasionally experience problems. But, like riding a bike, most of the problems you'll have with the Internet will occur during the first couple of sessions. After that, it'll seem easy and natural—like riding a bike.

GOING ONLINE YOUR FIRST TIME

Going online the first time can be both exciting and harrowing. As we've mentioned before, don't expect everything to work perfectly the first time. Your communications software might not be able to find your modem. The phone number for your online service or ISP may have changed (or be busy). A hundred things can conspire to keep you off the Internet, but once you have figured out the problems and finally log on, you can expect clear sailing (more or less) from then on. So don't think that every time you use the Internet will be like the first time, it won't be. (And it shouldn't be either—if it is, you need to get help from your online server or ISP's customer support staff.)

SETTING UP, SIGNING UP, LOGGING ON

By now you've installed and set up your communications software and application software (browser) as described earlier. You've got an account with an online service or ISP, which means you have an account name or e-mail address and a local phone number to call to log on to the Internet. Depending on which online service or browser you use, you'll simply click on the online service or browser's icon to start it (just like you would start, say, a word processor).

After the software loads, it should direct the modem to dial the online service or ISP's local access number. Then, depending on what software you are using and which service you subscribe to, the software will begin a "handshaking" ritual with the online service or ISP. Your modem and the service's modem will exchange signals to verify the quality of the telephone connection and make adjustments. They'll share information as to what transmission speed is desired. Once the modems are satisfied that a physical connection is possible, your application software will attempt to log on to the ISP or online service. You may be asked for your account name (or e-mail address) and a password. After that information is verified, you are in!

If you are using an online service, you'll go directly to their welcome screen. There, you can make selections from a menu of options. One of those options will be to connect to the Internet. If you are using an ISP, you will be automatically connected directly to the Internet. In either case, your browser will begin loading its default home page. It's called a default home page because the particular home page it loads is preset at the factory. (You can and probably should change this home page to a place you'd rather start from
—see Chapter 7.)

At this point, you can check for any e-mail messages. You'll probably have at least one already—a welcome message from the online service or ISP.

Also at this point, you can use your browser to go anywhere on the Web by simply opening a Browser window and typing in an address. (See pages 119 through 291 for lists of farming-related Web addresses, or see Chapter 11, *Search Engines*, for information about how to search for things on the Internet.)

As you'll quickly learn, Web addresses are called Uniform Resource Locators or URLs and are usually preceded with **http://** as in **http://www.rural.org**. The first part (http://) simply says that you are expecting information in hypertext transmission protocol (http). On most browsers, you no longer have to type all that http:// stuff in; the browser will add it automatically if you don't. You can try this by opening the browser window and just typing a URL such as **www.microsoft.com**.

Some versions of Netscape will add http://, www. and .com to microsoft to make the complete (and correct) address "http://www.microsoft.com." Although this feature is very useful, it can be somewhat confusing, since early versions of Netscape and most other browsers do not do this (Some browsers only add the www. and not the .com.). Netscape only adds the ending .com (not .edu, .org, and other endings). For example, if you just type in 'rural' to get to http://www.rural.org, Netscape will take you to http://www.rural.com. Unfortunately, that's not the same address as http://www.rural.org. For addresses which don't begin with www. or end in .com, you will need to type in a full address such as www.rural.org. If you have problems determining what shortcuts that your browser has built-in, don't worry. Remember you can always just type the entire URL into the browser.

At this point, you are ready to set up and learn to use e-mail, newsgroups, the Web and more. That's what the next few chapters are about.

When you've finished, you simply quit the application program, which will log you off the Net. (Some communications programs may have to be told you are ready to hang up—see the appendices for more information.) At this point, congratulations! You're now part of the worldwide Internet community.

SECURITY, VIRUSES, AND PRIVACY ISSUES

Netscape Navigator's "broken" key (top picture) means that your Internet connection isn't secure. The "solid" key (bottom picture) means it's (probably) safe to send credit card numbers and other personal information.

Forget what you've heard on TV about the problem of the Internet not being very secure. It isn't very secure, but that is not a problem to most people, most of the time. True, I wouldn't use the Internet to send highly confidential information, such as medical or financial records, unless that information was encrypted first. Fortunately, I've never had to send this kind of sensitive information over the Internet, but if I needed to, there are encryption programs built in to most Web browsers.

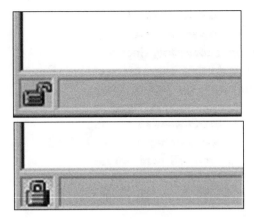

Microsoft's Internet Explorer uses a lock to indicate when your Internet connection is secure. The "open" lock means your connection is not secure. The "closed" lock means it's (probably) safe to send credit card numbers and other personal information.

As a rule of thumb, if you feel comfortable saying something on a cordless or cellular phone (or writing it on a post card), you can probably send it over the Internet without much danger. Your e-mail letter to Uncle Larry describing your recent vacation probably isn't what Internet snoops are looking for. Face it, people can probably find out more about you by going through your trash than by secretly reading your e-mail. The real problem comes from giving out credit card numbers over the Internet, which you might do to order a magazine or online newspaper, for example.

Today, most reputable companies doing business online handle such transactions in an encrypted mode. You can tell if your information is being encrypted by looking at the little key symbol at the bottom left-hand corner of most browsers' screens. In Netscape Communicator, look at the bottom of the active window for a lock. If the lock is open, then you don't have a secure communications link established, and an Internet snoop (with some rather sophisticated equipment and lots of free time) could, in theory, obtain your credit card number (if you typed it in) and play havoc with your credit rating. If the lock is locked, then you have a secure (encrypted) communications link and can safely use your credit card or exchange other sensitive information. Explorer works much the same way, except you'll see a lock when the link is secure and nothing when it isn't. See Netscape (or Explorer) Online Help for more information.

Of course, the encryption systems used by browsers can probably be broken, but it's not going to be easy to do. In one famous case of breaking a browser's encryption system, several (very smart) college students used about $100,000 worth of computers and lots of free time to break in. Considering the trouble and expense involved, it's not likely that someone will go to those lengths just to get your credit card number, unless your credit limit is really high (mine's not, so I don't have much to worry about). Again, the average user, exercising prudence and common sense, is probably not going to have to worry too much about security on the Internet.

Viruses are another supposed problem. Viruses are tiny programs (written by antisocial types) that are designed to screw up your computer by damaging data and files. Again, if you are not on a network at work or school (which is where most of this nonsense takes place), viruses should not be a problem to you if you take the following precautions. Since viruses are little programs, you can't get a virus unless you download and run software on your computer. This means that you can't get a virus by reading e-mail or looking at a Web site. However, it is possible get a virus by opening some word processor or spreadsheet files that contain *macros*. Macros are little programs that can be embedded in some word processor or spreadsheet files to handle repetitive tasks. Since macros are really programs and not just text or numbers, they can contain commands that could do things to your computer that you don't want. It was reported recently that one prankster was e-mailing to people at random a Word file entitled "AOL4FREE" that contained a macro virus. Supposedly, the text file told the reader how to get AOL for free while the macro did it's dirty work. To (virtually) ensure that a word processing or spreadsheet file does not contain an "infected" macro, don't open one of these files unless you know the sender.

Most companies that offer software over the Internet take great pains to insure that the software you download is virus free. Furthermore, you could always install a utility program that checks for and eliminates viruses from your computer.

At this writing, the version of Netscape Navigator or Internet Explorer that you are most likely to have uses a 40-bit encryption system. While this is usually good enough for credit card type transactions, some people do not consider it good enough for "big dollar" transactions or highly confidential information.

For example, if you trade commodities or stock online, your broker may require that you obtain a special 128-bit version of Netscape Navigator or Internet Explorer to insure that your transactions are more secure. These versions are available only to U.S. and Canadian citizens, but you can download them from the Web. For Netscape, the address is www.netscape.com. For Internet Explorer, the Web address is www.microsoft.com/ie. Look for the 128-bit encryption version of the browser you want and download it. You will be asked to fill out an electronic form certifying that you are a U.S. or Canadian citizen.

CHAPTER 3
E-MAIL

E-mail stands for electronic mail, and it is one of the most useful, flexible ways of communication available. You can use e-mail to send a message to anyone who is on the Internet or who subscribes to an online service like AOL. You can send e-mail to one person, or you can "broadcast" to many. With most e-mail programs, you can attach a file to an e-mail message—the file could be another document, a picture, or even a sound clip.

E-mail is very fast. An e-mail message can travel to the other side of the world in seconds. E-mail is inexpensive, too. Most commercial Internet access providers allow users to send "unlimited" amounts of e-mail as part of their basic monthly charge. While many Internet providers have established some upper limit on the amount of e-mail any one person can send in any one month, this amount is usually so large that it's unlikely you will ever exceed it. (I can't recall ever having gone "over the limit," and I send a lot of e-mail.) Therefore, being "free," e-mail is cheaper than regular mail (which is called "snail mail" by the Internet community).

E-mail is ready when you are—and it will wait for you until you're ready to read it. Your e-mail is stored in your mailbox on your ISP or online service's computer. When you log on to your Internet service and check for e-mail, your messages will be waiting for you. Now for the best part: if you send a message to someone in another time zone, you never have to worry about whether you're calling at a bad time. This is especially valuable for farmers who never keep "banker's hours."

E-mail goes to your electronic Internet address, so it can follow you wherever you go. For example, if you use a portable computer with a modem, you can plug into a phone line anywhere in the world, dial your Internet service, and check your e-mail.

In addition to using e-mail for one-to-one communications, e-mail is also used to receive messages from mailing lists. There are several farming-related mailing lists you can subscribe to—for free—and automatically receive information from time to time about a farming-related topic. See Chapter 4, *Mailing Lists*, for more information.

E-mail is also used to post messages from you to a newsgroup while another type of program—called a newsgroup *reader*—receives and processes messages to you from the newsgroup. (Newsgroups are groups of people on the Internet who hold discussions and exchange information about a particular topic—see Chapter 5, *Newsgroups*, for more information.)

If you have to watch the clock while you're online because (A) you have to make a long-distance call to connect to your Internet service or (B) your Internet Service Provider has expensive rates and charges by the minute, you can save time and money by writing and reading e-mail while you are offline (not connected to the Internet). To do this, write your e-mail messages before you connect to the Internet. After the messages are complete, connect to the Internet and send the messages quickly, and log off (disconnect). Likewise, you don't have to be online to read e-mail messages. After you go online and retrieve your e-mail messages, simply log off and read them at your leisure without watching the clock.

E-MAIL SOFTWARE

If an online service, like America Online, is your Internet provider, you'll find an e-mail program already built in to their software package, so you won't need a separate program. But if you are using a true Internet Service Provider, such as NetCom, MCInet, etc., you will need an e-mail program of some sort. As discussed in Chapter 2, many ISPs provide free software as part of a starter kit, so you may not have to obtain an e-mail program on your own. (The ISP's software package may include a stand-alone e-mail program like Eudora—as well as other stand-alone programs for reading newsgroups and doing other things--and will almost certainly include a Web browser—usually either Netscape or Explorer.)

While you could install and use a stand-alone e-mail program like Eudora, you are probably also going to install and use a Web browser like Netscape or Internet Explorer. Today's Web browsers now include built-in software to handle e-mail and other Internet services. Once the Web browser is installed and properly set up, it will handle all of your e-mail tasks (as well as browse the Web) without further ado. Thus, there's no need to install separate programs to handle e-mail, newsgroups, etc. (See Chapter 7 for more information.)

Our advice is to install one of the newer versions of Netscape or Internet Explorer, which handle all of these services, and be done with it. However, some of our colleagues have an alternative view. When we noticed that some of them still used Eudora, even though they also had Netscape with its built-in e-mail system, we asked them why they needed two e-mail systems. The answer was that Netscape is a large program that takes a while to load and run. They use Eudora when they are in a hurry and want to quickly check their e-mail. Another reason was that Eudora was a small program and could be run along with a word processing program on computers that didn't have a lot of memory.

E-MAIL ADDRESSES

If you think about it, your parents didn't have a lot of choice when they named you—if they followed Western tradition. They may have had complete freedom to choose your first and middle names (depending on which relatives they needed to please), but your last name is probably the same as your father's. After your parents chose your first and middle names, these were added to your father's last name and recorded on a birth certificate. At that point, you are who you are.

E-mail addresses work much the same way. The "last name" in an e-mail address is called a *domain name*. The domain name is usually constructed from the Internet provider's name, a dot, and a three-letter code. The code identifies what type of organization is supplying your Internet connection. Here is a list of the most common types:

.com	**businesses**
.edu	**schools and research facilities**
.gov	**government agencies**
.mil	**military facilities**
.net	**major Internet access providers**
.org	**organizations (non-profits and service groups)**

For example, the domain name for the University of Kentucky is **uky.edu**, while the domain name for the Tennessee Valley Authority is **tva.gov**.

A complete e-mail address usually consists of your name (or parts of it, anyway) and your ISP's domain name. They are combined with the @ symbol, as in **yrname@aol.com**. If NetCom is your Internet provider, then your e-mail address probably ends in **@netcom.com** (pronounced "at netcom dot com"). If the Microsoft Network is your provider, then the last part of your e-mail address will be **@msn.com**, and so on.

When you first sign up with an Internet provider, you will be asked to choose an e-mail address. Within reason, you'll get to pick the first part of the e-mail address. There are character limitations, so you may have to abbreviate a little such as **jbarker** instead of **jonathanbarker**. (Note than there are no spaces in e-mail addresses.)

Another limitation is that no two people can have the same e-mail address. There can be a **jsmith@msn.net** and a **jsmith@aol.com**, but there can't be two people with the exact same e-mail address. If you use an ISP who has lots of customers (and this is especially true for online services), there may already be several users with your name when you sign up, so an e-mail address may have to be part name and part number. It's not usual to see something like **jsmith55@aol.com**. (Please don't send e-mail to this Mr. or Ms. Smith—if he/she exists. We made up these addresses as an example.)

Another important point about e-mail is that e-mail addresses are not portable and Internet providers don't have to forward your e-mail once you cancel your subscription. For example, say that you sign up with AOL and your e-mail address becomes **jsmith55@aol.com**. Now, several months later, some other ISP offers you a better deal and you take it. You may take the deal, but you won't be taking your e-mail address. Your new Internet provider may let you keep the **jsmith55** part, but they can't let you have the **@aol.com** ending. Instead, your new e-mail address would become **jsmith55@newisp.com**. Since these addresses are different, e-mail sent to **jsmith55@aol.com** won't get to **jsmith55@newisp.com**. The point is to make sure you are happy with your Internet provider before giving your e-mail address to too many people, printing it on stationery, or painting it on the side of your truck.

E-MAIL PASSWORDS

In addition to selecting an e-mail address (or at least the first part) you will be asked to select a password. When you log on (connect) to your Internet service and ask to see your e-mail, the system will ask for your password. Without it, you won't get your e-mail. In selecting a password, pick something that you are certain you'll remember or write it on a piece of paper and tape it to the bottom of your mouse pad.

Most Internet providers have particular requirements for passwords. They must contain at least a certain number of characters (usually five is the minimum), and they can't be over a certain length (usually 10 characters is the maximum). For added security, some ISPs require that passwords contain both letters and numbers. For example, my password is. . . which reminds me, never give your password to others unless you want them to be able to read your e-mail and send messages while pretending to be you.

You can pick just about anything you want for your e-mail address (subject to certain limitations). This address represents YOU to friends and relatives and to universities, government agencies, and companies. What may sound cute today might stick in your throat tomorrow when you have to tell some guy at USDA that your e-mail address is bunnyhutch@aol.com. Pick an e-mail address that truly represents you. As Roger Ailes says, You are the Message.

WHERE DOES E-MAIL GO WHEN YOU SEND IT?

When you create an e-mail message and send it to someone else, it goes out to a "mail server." (A "server" is a computer on a network that performs network management tasks, such as storing programs and routing information from one computer to others on the network.) The mail server determines the best route for your message and passes the mail to the next closest server. Your message may be passed through many servers before it arrives at the final recipient's mail server, but usually it takes only a few seconds or minutes at most.

Online services, like CompuServe and AOL, have their own networks of mail servers all over the world. These private networks are connected to the Internet, and if a message is going to a computer not on their network, it can be passed to the Internet. All this occurs very rapidly; most e-mail messages are delivered in a matter of seconds. But sometimes e-mail is not delivered perfectly intact when going to and from online services to the Internet.

If everyone in the family has his or her own e-mail address, then everyone will have to check his or her e-mail (every day or so) to see if they've received any e-mail. That can add up to a lot of online time spent by everyone just checking for e-mail. For many families, it may be better to have one e-mail address for the entire family and maybe another to represent the farm. In some cases, just one will suffice.

USING E-MAIL

Beneath the variations in appearance, you'll find that most e-mail programs work basically the same way. Virtually all e-mail programs allow you to perform these basic tasks:

- **Send mail**
- **Read mail**
- **Print mail**
- **Keep an "address book" of your correspondents**
- **Attach other files (like word processing files) to an e-mail message**

The following instructions are for "generic" versions of Netscape and Explorer. A particular version of a browser has a "shelf life" of about six months; after that, the software makers "improve" their offerings by moving a few buttons around and changing a few features. Also, the versions produced for Windows 98, Windows 95, Windows 3.1 and the Mac are all slightly different. We've tried to point out those differences where they might make you stumble, but with four major operating systems each getting a new browser every four to six months, the permutations are endless. As a result, the particular version of Netscape or Internet Explorer you are using may work differently from the examples below. However, both programs have built-in help features (online help) that should take care of most of the problems you run in to and can teach you to use other more advanced features as well. Most online services have online help, too.

READING AND SENDING E-MAIL

First, start the Web browser (Netscape or Internet Explorer). Starting the browser should trigger a chain of events that will eventually connect you to your ISP. As it comes from the factory, your browser probably starts by trying to load a home page from the Web (see Chapter 7). If you are impatient, you can click the **Stop** button and go directly to the e-mail system without having to wait for the Web page to load.

You can set most versions of Netscape to start with e-mail (or news) instead of starting with the Web browser. First, pull down the Options menu and select General Preferences. Now, you'll see a section called "On Startup

Launch" that has three buttons. By clicking on one of the buttons, you can tell Netscape to start with the Web browser (the default) or Mail or News.

Microsoft's Internet Explorer uses a different method (at this writing) for accessing e-mail and news. In Windows 95, select **Programs: Internet Mail** (or **Programs: Microsoft Outlook Express**) to start the e-mail software without first running the Web browser (Explorer). On the Mac, simply click on **the Internet Mail and News** icon in the Microsoft Internet Applications folder instead of launching the Explorer browser.

READING E-MAIL USING NETSCAPE

Pull down the **Window** menu (or the **Communicator** menu, if you are using Netscape Communicator) and select **Netscape Mail** (or **Messenger Mailbox**). Depending on how you originally set up your copy of Netscape, you may be asked for a password, and you may have to click on **Get Mail** to get things started.

Newer versions of Netscape have a tool bar outside the browser window where you can choose between the Web browser (Navigator), e-mail or newsgroups (called Discussions).

A typical e-mail message using Netscape Communicator.

After you click on **Get Mail**, Netscape will ask your ISP's server if you have any mail, and it will retrieve it for you if you do. By looking at the bottom of the Mail window, you can see how many messages you have and follow the progress as e-mail messages are received.

The Netscape e-mail window has many features. The Folder window on the left shows you how many messages are in your "Inbox." The new e-mail messages are listed as Unread. The ones you've already read are listed as Total. From time to time you can move old e-mail messages to the Trash but you'll have to manually "carry out the Trash" by pulling down the **File** menu and selecting **Empty the Trash**. (The location of the **Empty the Trash** command may vary depending on which version of Netscape you have.)

The Subject window on the right tells you what each message is about, who sent it and when. (The bar between these two windows can be moved back and forth to let you see more information. See Netscape's Online Help for instructions.)

The large window at the bottom contains the text of the first message. Other messages can be viewed by clicking on them in the Subject window (or by sliding the Subject scroll bar). Use the scroll bar on the right side of the message itself to move from beginning to end on long messages.

At this point, you can use the command buttons at the top of the Mail window to reply to the sender (Re: Mail) or to reply to everyone who received the same message (Re: All). You can also forward the message to another person (Forward), print the message on your printer (Print), or trash it (Delete). All messages are saved to your hard drive until you Delete them. Even then, they are still stored in an e-mail trash folder until you empty it.

You can also save a message permanently by pulling down on Netscape's **File** menu and selecting **Save As**. Or you can use your mouse to highlight the text (or selected portions) and copy the highlighted text to a word processor using cut and paste techniques—see your word processor's instructions for more information. You can set up most e-mail systems to store e-mail messages either on your computer or on your ISP's server. It's probably a good idea to store your e-mail on your own computer rather than on someone else's.

If you're paying long-distance charges to connect to the Internet, don't print while you're online. It takes too long and can be costly. Instead, save the message to disk (your e-mail program probably does this automatically) and print after you've gone offline. Saving to your disk drive is much faster than printing to your printer. See Appendix B for more information.

SENDING E-MAIL USING NETSCAPE

To send an e-mail message, simply click on the **To: Mail** (or **New Msg**) button. Now, type in the recipient's e-mail address *exactly*. Next, add a subject (and a cc recipient if you want someone else to get a copy). Then click on the large empty message window and type your message.

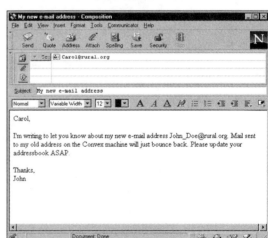

E-mail messages require an e-mail address, a subject, and, of course, a message. You can also attach other files such as spreadsheets, documents, and pictures.

You may want to use your word processor to compose long messages which you can import to the e-mail message window using cut and paste techniques. Alternatively, you can write a message using your word processor and attach the word processing file to an e-mail message. The only problem with this approach is that the recipient must have a compatible word processor that can open and read the file you've attached. If you are not sure, save the word processing file as a plain text file (ASCII), and they should be able to read it. This compatibility issue also applies to other kinds of files such as graphics and spreadsheets—make sure the recipient has a similar program that can open the file before you attach and send it.

To attach a file, click the **Attach** button. Now, select **Attach File** and go through the folders on your computer until you've found the file you want to attach. You can attach more than one file. When you're finished, click on **Done**. After you've finished addressing and composing the message and selecting any attachments, simply click **Send Now**, and it's gone.

E-mail addresses you use often are most easily kept in an electronic Address book that is included with most e-mail systems. Once you've added an address to your Address book, you won't have to type it in again. How the address book feature works varies from one version of Netscape to another, but in general, here's how to do it:

To open the Address book, click on Address from the To: Mail window (or you can pull down the Window menu and select Address). To add a user, pull down the Item menu and select Add User.

If you are replying to an e-mail someone sent you, you can simply click on Re: Mail (which stands for reply mail) and Netscape will automatically enter the e-mail address (from the message to which you are replying). Sometimes it is necessary to quote all or part of a message that's been sent to you. You can use the Quote button for that. However, unless the recipient has a short-term memory problem, it is not a good practice to include the entire text of a received message in every reply.

Although your e-mail address is transmitted with every message you send, it's a good practice to add what is known as a signature to your e-mail messages. A signature is a small file than can contain your full name and address, your phone and fax numbers, and other contact information. See Netscape's online handbook for more information.

READING E-MAIL USING EXPLORER

E-mail and Newsgroups work somewhat differently in Explorer, so don't be surprised to also see News when you're expecting just to see Mail. Also, different versions of Explorer's Mail and News systems vary in how you read e-mail. Some (newer) versions of Explorer call the Mail and News system "Microsoft Outlook Express." You may need to check the built-in help system for information about the particular version of Explorer you have. Nonetheless, the following generalized instructions should help you read e-mail.

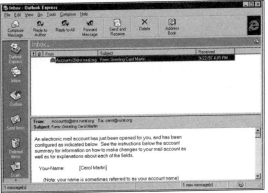

A typical e-mail message as received through Microsoft's Internet Explorer.

In Windows 95, click on the **Mail** icon and select **Read Mail** (or select **Go** from the pull-down menu and select **Read Mail**).

The Macintosh versions of Explorer may use either Eudora Lite or Microsoft's Mail and News as the e-mail system. Although they are very similar in function, they do differ a little in how they work. If your version of Internet Explorer uses Eudora Lite, you can send e-mail from Internet Explorer (by clicking on the "speeding envelope" icon), but you'll have to run Eudora Lite in order to receive (read) e-mail. See Internet Explorer help for more information.

If your version of Internet Explorer uses Mail and News as the e-mail system, then you'll be able to access e-mail directly from Explorer's main window. To do this, click on the mailbox icon. Now the menu bar will change, offering several options. Pull down the **Mail** menu and click on **Send/Receive**. Alternatively, if you are already in the Mail window, you can also click on the **Send/Receive** icon.

Next, Explorer will ask your ISP's server if you have any mail, and if you do, it will retrieve it for you. By looking at the bottom of the Mail window, you can see how many messages you have and follow the progress as each e-mail message is retrieved. The window on the left shows several folders set up as In and Out boxes as well as Help. The top of the right window shows the subject of each message, who sent it, and when.

The large window at the bottom contains the text of the first message. Other messages can be viewed by clicking on them in the Subject window. Use the scroll bar on the right side of the message to move from beginning to end in long messages.

At this point, you can use the command buttons at the top of the Mail window to reply to the sender (Reply to Author) or to reply to everyone who received the same message (Reply to All). You can also forward the message to another person (Forward) or trash it (Delete). All messages are saved to your hard drive until you Delete them. You can also save the message as a text file by pulling down on Explorer's **File** menu and selecting **Save As**, or you could use your mouse to highlight the text (or selected portions) and copy the text to a word processor using cut and paste techniques—see your word processor's instructions for more information.

You can print from Mail just like you would from any other program by double-clicking on the message, pulling down the **File** menu bar and selecting **Print**.

SENDING E-MAIL USING EXPLORER

To send an e-mail message simply click on **New Message** (or the "speedy envelope" icon, depending on which version of Explorer you have). You'll need to know the recipient's e-mail address. E-mail addresses are most easily kept in an electronic *address book*. Once you've entered an address in the Address Book, simply click on the icon (which looks like a Rolodex card) and select a recipient from the list.

E-mail messages require an e-mail address, a subject, and, of course, a message. You can also attach other files such as spreadsheets, documents, and pictures.

Next, add a subject (and a cc and bcc recipient if you want to send a copy to someone else). Then click on the large, empty message window and type your message.

You may want to use your word processor to compose long messages which you can import to the e-mail message window using cut and paste techniques. Alternatively, you can write a message using your word processor and attach the word processing file to an e-mail message. The only problem with this approach is that the recipient must have a compatible word processor that can open and read the file you have attached. If you are not sure, save the word processing file as a plain text file (ASCII), and they should be able to read it. This compatibility issue also applies to other kinds of files, such as graphics and spread sheets—make sure the recipient has a similar program that can open the file before you attach and send it.

After you've finished addressing and composing the message, simply click on Send/Receive, and it's gone.

Most e-mail systems let you set up a signature file that can be attached to the bottom of your e-mail messages. This file can contain just your name, or it can include your e-mail and post office addresses and phone and fax numbers, too. Signature files are a good way to ensure that a recipient can contact you by ways other than e-mail, and you will only have to click on the Signature button to include this important information in an e-mail message. To set up a signature file in Netscape, you'll have to use a text editor (word processor). Write whatever information you want to include (usually just a few lines) and save the file. Then run Netscape. Pull down the Options menu and select Mail and News Preferences. Now select Identity. (For Netscape Communicator, select Edit:Preferences, then open Mail & Groups and select Identity.) Now, click on the File button (under the Signature File) and then click on the Browse (or Locate) button to show Netscape which file you want to use as your signature. This signature file will be attached to the end of every e-mail message you send (unless you go back to the Identity window and select None as the signature file).

In Explorer, pull down the Edit menu and choose Preferences. Click on the Open Internet Config button and you'll get a selection of things you can personalize. Click on the Personal button and type your name, address and phone numbers in the Signature box. When you are sending an e-mail message and want to include this information, simply click on the Signature icon, and it will be added automatically to the e-mail message.

SENDING AND RECEIVING E-MAIL USING ONLINE SERVICES

Most online services tell you if you have e-mail as soon as you log in (connect). "You've got mail!" However, how you actually get to your e-mail and read it depends on which service you are using and which version of the online service's software you have installed. As confusing as this may sound, most are simple to use, and most have online help if you get into trouble.

Take America Online for example. To read or send e-mail, you pull down on the **Mail** menu bar. You now have the option of writing a message (**Compose Mail**), reading new messages (**Read New Mail**), or reading old messages you've already read. You can also create and make changes to an address book (**Edit Address Book**). If you get stuck, most online services have either online help or help applications built right into the software. However, this help file is often so large that it's usually not included on the distribution disk and may have to be downloaded from the online service the first time you use it. However, once you have it, you'll be able to use it without having to download it again. (Some people use the help system only once. Since it takes a while to download, they mistakenly believe that it's going to take a long time to use every time they click on **Help**. This download only happens the first time and may not happen at all if you installed the software using the CD-ROM version instead of the much smaller floppy disk.)

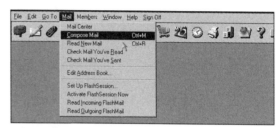

America Online's "user friendly" interface makes it easy to send and receive e-mail.

To send an e-mail message, simply pull down the Mail menu bar and select Compose Mail. To get your e-mail, select Read New Mail.

Online services like AOL and CompuServe vary in how they handle attachments (word processing documents, spreadsheets, pictures and other files) sent via e-mail. Because there's no universal standard (yet) for attaching files to e-mail messages, sending attached files between online services or between the plain old Internet and an online service may not always work. Of course, there is usually no problem going from AOL to AOL, or CompuServe to CompuServe, or from one ISP to another ISP. Also, sending e-mail from any online service to the plain old Internet (ISP) usually always works. The problem (if you experience one) usually happens when you try to send e-mail (with attachments) from the plain old Internet (ISP) to an online service subscriber.

CHAPTER 4
MAILING LISTS

If your mailbox is stuffed regularly with sale flyers from the co-op, church bulletins, and extension service newsletters, it is because you are on several mailing lists. You are on these lists because you have an interest in the subject of the mailing. You shop at the co-op, you attend that church, and at one time or another, the county agent got your name and address and started sending you a newsletter. While these examples all represent "paper" mail, the Internet has its own electronic version of the mailing list. Obviously, with the Internet version, you get e-mail instead of paper envelopes, but the principle is the same. Your name and address (e-mail address) are stored on someone's mailing list, and when they have an announcement to make to the "list," they e-mail the announcement to
everyone on the list.

Today, there are more than 30,000 public mailing lists on the Internet. In the early days of the Internet, mailing lists were often called "listservs" and some still are. But regardless of what they are called, they simply are a way to send e-mail *automatically* to lots of people on a regular basis.

Most Internet mailing lists are organized around an area of interest like farming or a particular topic like beef. A couple of beef related mailing lists are Beef-L and BeefToday-L. (See page 203 for other beef related mailing lists and information on subscribing to the lists.)

There are literally thousands of mailing lists available from the Internet. Some lists are simply pipelines for information and do not encourage member-to-member discussions. The majority of mailing lists, however, encourage the active exchange of information and ideas among list members.

To get on a mailing list, you must have an e-mail address. You subscribe to a mailing list by e-mailing a subscription request to the mailing list's address along with a short message that tells the list manager you want to subscribe—see page 52. Once you're "on" the mailing list, any messages sent to the mailing list are automatically e-mailed to you. Subscriptions to mailing lists are almost always free, but there are some exceptions (such as high-value stock market and commodity information and the like).

Some mailing lists don't have much activity and won't send very many messages, but others do and some can quickly fill your e-mail box. So, join only one or two mailing lists that relate directly to your farming operation and see if you like them—you don't want to join too many and receive an unmanageable amount of e-mail. See page 215 for farming-related mailing lists that you can subscribe to or check under specific headings (like dairy on page 206) to find topic-related mailing lists.

Mailing lists fall roughly into the following categories:
- **Electronic newsletters**
- **Electronic magazines**
- **Public debate and discussion forums**
- **News and information dissemination lists**
- **Small clubs where friends share information**
- **Personal information distribution lists**

With most mailing lists, you are simply on the receiving end. You get what they send you (such as a newsletter), and there is little opportunity or need to respond. With other mailing lists, subscribers can post messages to the other subscribers on the list. Mailing lists that allow subscribers to post messages usually have a "moderator" who reviews all messages before distributing them to the list. This process is supposed to prevent the discussion from straying off the topic, but most mailing lists are unmoderated or are moderated by "smart" computer programs instead of humans.

Messages come fully identified as to the author and address, so you can respond either to the author or to the whole list. If you don't want to broadcast your response to the whole mailing list, you should send your e-mail to the author directly. In fact, you should never send a message intended for one person to the whole list—that's considered a no-no in Internet Land.

Like most online services, America Online has thousands of mailing lists to choose from.

FINDING INTERNET MAILING LISTS

The authors of this book have already assembled a ready-to-use list of farming-related mailing lists (page 215), which is the easiest way to find mailing lists. However, you can search for other mailing lists on your own by using the Internet's search engines (see Chapter 11) or by using the following instructions:

1. Start your Web browser (see Chapter 7).
2. After the home page loads, move the mouse's pointer to the **Go To** box (if you're using Netscape) or the **Address** box (if you're using Internet Explorer). Next, erase whatever address is in the box and type **http://www.rural.org** as the new address and press enter. In a few seconds, the Rural Studies Home Page will load. Now, click on Farmer's Guide Updates. Among the many choices will be Mailing Lists. Click on Mailing Lists for an up-to-date list.

ONLINE SERVICES' MAILING LISTS

Any online service that can connect you to the Internet (such as AOL or MSN) automatically gives you access to just about any mailing list on the Net. However, some online services often have their own mailing lists which may or may not be available to non-subscribers who are on the Internet itself. Most online services automate the search process, making it very easy to find mailing lists.

For example, to locate a mailing list on AOL:

1. On the main menu, click **Keyword** (or pull down the **Go To** bar and select **Keyword**).
2. Type **Mailing Lists** as the keywords and press enter.
3. Now, you can browse AOL's mailing list directory or search for mailing lists by subject. Once you've found a mailing list you like, you can subscribe to it via e-mail.

America Online lets you browse through a list of mailing lists or you can search for a mailing list by subject.

SUBSCRIBING TO MAILING LISTS

Mailing lists are not consistent in how they want you to provide subscribing information, but almost all of them have instructions about how to prepare an e-mail message that will get you on their mailing list. For the most part, you simply send the list manager an e-mail message with the word "subscribe." But some want that word in the body of the message and some want it as the subject.

Some mailing lists can generate lots of e-mail and clog your mailbox with stuff you're too busy to read. Be sure to read your Internet provider's guidelines about how your e-mail is managed. Sometimes they'll throw e-mail away after a certain date or if your e-mail file gets too large.

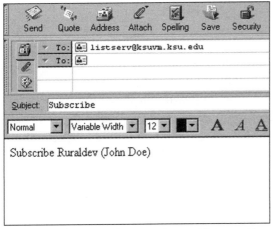

A typical mailing list "subscription" using Netscape. Be sure not to lose the instructions about "unsubscribing" or you'll have a hard time getting off the mailing list.

TO SUBSCRIBE TO A MAILING LIST:

1. If possible, try to find out something about the list before you subscribe. In particular, be sure to note the *subscription address*—this may be a different address than the one to which you'll post messages.

2. E-mail your name and any other requested information to the subscription address. The list's subscribing information will explain exactly how to do this. Usually it involves sending a message that contains the word "subscribe" (your e-mail name is already there, in the header of the message). However, some lists want you to put the word "subscribe" in the subject line instead of in the message itself.

3. Within a day or so, you will probably get a confirmation note. You may be asked to send in some additional information about your interests. You will also be given the posting address (where messages should be sent) and information about posting messages to the list.

TO UNSUBSCRIBE TO A MAILING LIST:

You can get off a mailing list at any time—in theory. To unsubscribe, you usually follow the same process as you did to get on the mailing list, except that you e-mail the subscription address a message that you want to "unsubscribe."
However, some mailing lists have different instructions for "unsubscribing." For this reason, keep the information about subscribing and unsubscribing in case you want to get off the list.

It's a good idea to save the original subscription instructions so that you will be able to unsubscribe. Getting off a mailing list is easy if you've kept this information, and it can be very troublesome if you haven't and don't know how to get off.

CHAPTER 5
NEWSGROUPS

Newsgroups are discussion forums organized around a particular interest, issue, or activity. Newsgroups differ from mailing lists in two important ways. First, as you will recall from Chapter 4, a mailing list is an automated system that periodically sends separate e-mail messages to everyone who "subscribes" to that particular list. Thus, subscribing information (your e-mail address) is kept on a computer somewhere out there in cyberspace. And, as described in Chapter 4, you must send in a "subscription" request to join the mailing list.

Newsgroups, on the other hand, are continuing forums that you check in with every once in a while—when *you* want to. You are not on the newsgroup's "list," instead, the names of the newsgroups you participate in (if any) are stored on your computer in a preferences file maintained by your Web browser (or stand-alone newsreader program). Another difference is that a mailing list's information consists of separate e-mail messages that are sent to your e-mail address whenever a message is posted to the mailing list.

A newsgroup, consists of a single (sometimes very large) file that is stored on a computer called a *news server*. This one file contains all the "postings" or messages that have been sent to the newsgroup. Instead of getting these messages sent to you one by one, *you* decide when you want to read your newsgroups— it's not automatic.

To read a newsgroup, you'll need a special program called a "newsreader," although most browsers like Netscape and Internet Explorer now have built-in newsreaders.

When you want to read a newsgroup, your Web browser (or stand-alone newsreader program) contacts the news server where the newsgroup is kept and downloads the whole file (or if the newsgroup file contains a large number of postings, the browser may just get a list of everything that's in the newsgroup's file and you can download individual postings by clicking only on the ones you want to read). Although mailing lists and newsgroups operate quite differently, they are both used to share information about a particular subject—mailing lists use e-mail and are automatic, newsgroups use a newsreader and you have to activate it.

Newsgroups have several practical advantages over mailing lists: they don't clog up your e-mail in box, discussions are more "conversational," and you can quit simply by not reading them anymore.

You can use newsgroups to share expertise and information, ask questions, or debate current issues. There are over 30,000 newsgroups on the Internet, and the number grows every day.

The newsgroup format—the way they look and work—evolved from UNIX-based message posting and conferencing programs used by computer science and graduate students in the 1970s. This network of newsgroups was originally called UseNet. You may still see newsgroups referred to as "UseNet newsgroups," although most modern browsers simply refer to them as *News*.

Newsgroups are maintained by the people who created them. Some newsgroups are *unmoderated*—no one looks at the messages before they are sent to the newsgroup file. As a result, you'll find that the content of unmoderated newsgroups varies widely—if not wildly. (Newsgroups sponsored by online services *are* monitored by the online service itself and are usually much tamer.) Other newsgroups are *moderated*, so that only messages approved by the newsgroup's administrator or moderator will be posted. But moderated newsgroups are the exception, not the rule. In most newsgroups, every message sent in (posted) ends up in the discussion (called a *thread*).

SOFTWARE FOR NEWSGROUPS

To read a newsgroup, you'll need a special software program called a "newsreader." Most Web browsers like Netscape and Internet Explorer have built-in newsreaders. (At this writing, Internet Explorer uses a News and Mail subsystem that's physically separate from the Web browser, but it's nicely integrated and is, of course, included with the Web browser.) The online services also have their own built-in newsreaders; America Online's is excellent.

SUBSCRIBING TO A NEWSGROUP

To subscribe to a newsgroup, you'll first have to identify the <u>name</u> of the newsgroup server to which you have access. Online services like America Online already know what newsgroup server you'll be using, so there's nothing to figure out. But if your Internet provider is an ISP, you will have to ask your ISP for the name of the news server you are assigned to and then let your browser's newsreader system know the news server's name. The name of your newsgroup server is probably "news" or "NNTP," but it could be something else. In any case, only your ISP knows for sure.

Once you've determined the name of the news server you'll be using, you have to set up (configure) your browser (or newsreader) to look for it. How you do this depends on what operating system you use, whose browser (or newsreader) you use, and where they've moved the configuration or preferences screens since we wrote this book. The latter is a real problem since even the "official" guides to Netscape and Microsoft's Internet Explorer don't seem to be able to keep up with these changes either. Nevertheless, the instructions below are for widely used versions of Netscape and Explorer, and should help, but may not match exactly the particular version you have. If that's the case, check with your ISP or use the built-in help systems that both Netscape and Explorer have (see Chapter 7).

CONFIGURING NETSCAPE:

Go to the **Options** pull-down window, select **Mail and News Preferences**, then select **Servers**. (With Communicator, select **Edit: Preferences** and select **Mail & Groups** and then **Groups Servers**.) On the screen you should see a little box marked "News (NNTP) Server." The place for the server's name may already say "news" (or "news.myisp.com") and this may be the correct name for your ISP's news server. If not, type in the correct server name. Also, right below is a box that tells the server how many messages you are willing to accept each time you read your newsgroups. The default is 500. You might consider changing this to some smaller number (50) until you have read the newsgroup and are certain it meets your needs. Otherwise, you could end up waiting a long time for 500 messages to download—messages you may decide you don't want after you've read one or two.

CONFIGURING INTERNET EXPLORER:

There are (at least) two ways to specify a newsgroup server in Microsoft's Internet Explorer. If you are using a program called "Internet Config" (which allows different Internet applications to share information) and you have specified the correct news server in Internet Config, you don't have to do anything else. Your ISP may have provided Internet Config as part of its starter kit. If not, check out Internet Explorer help for information about how to download a copy of Internet Config.

If you don't have Internet Config, you can simply click on either the **Mail** or **News** button, which activates the Mail and News subsystem. Then pull down the **News** bar and select **Options**. Now, select the **Server** bar and type the name of your ISP's news server in the Server Address box. If your version of Internet Explorer has Internet Config (and you don't want to use it to specify your news server), unselect the box that says "Get news server information from Internet Config" and enter the information manually.

With some versions of Internet Explorer, you click on the **Mail** button, then select **Read News**. If your browser has not yet been configured, it will ask for your name, your e-mail address and your ISP's news server. If your news server requires a user name and a password, you can specify them here also.

Once you have identified your news server and have configured your browser to look for it, you can now download a list of newsgroups contained on that server and pick the ones to which you want to subscribe. By the way, not all news servers provide access to all newsgroups—there are simply too many and some are obscure and not widely read. But most news servers give you access to tens of thousands of newsgroups, so you should be able to access all the ones we list in this book (see page 146).

HOW ARE NEWSGROUPS DIFFERENT FROM MAILING LISTS?

At first glance, newsgroups and mailing lists seem pretty similar. The main difference is in who controls the subscription. As we discussed in Chapter 4, to subscribe to a mailing list, you actually send in an electronic subscription request to the list's manager. Once you are on a mailing list, you must send in an electronic subscription cancellation message to stop them from sending any more messages.

Newsgroups are different in that the information about which newsgroups you subscribe to is stored on your computer, not someone else's. If you want to subscribe, you tell your computer to download the messages posted to that newsgroup. When you want to quit reading a newsgroup, you can deselect it from the list or simply just don't read it anymore. Another difference between mailing lists and newsgroups is that with a newsgroup, you have more control over what you view. If you are on a mailing list, you receive every e-mailed message that's posted to the list. And each message is a separate piece of e-mail. With a newsgroup, you can use your Web browser to scan through messages and select (and download) only the ones you want to read.

net tip: A message sent to a newsgroup is called a posting. An ongoing conversation about a single subject is called a thread. For example, someone might ask "Did you post anything to that thread about the farm provisions of the new tax code?"

SELECTING A NEWSGROUP

Since there may be more than 30,000 newsgroups on your ISP's news server, finding the particular news group you want can take time. The easiest way to find the names of newsgroups about farming-related subjects is to use the list on Page 146. Alternatively, you can search for them by using special newsgroup search engines like DejaNews **www.dejanews.com** (see Chapter 11).

After you've identified which newsgroups you want to read, you must let your newsreader know. The newsreader will remember which newsgroups you like, and the next time you want to read them you only have to click **Read My Newsgroups** (or some similar command).

Unlike most mailing lists, newsgroups usually encourage discussions among "members." You can see whether or not you like a newsgroup and want to join by reading the postings for awhile. Those who read without posting comments of their own are said to be "lurking." In fact, the vast majority of people who use newsgroups are lurkers who read the various "threads" and extract what's useful, without ever participating in the discussions. That's OK, though it's not as much fun. However, it is considered bad form to join a newsgroup and start posting messages before you have a "feel" for what's going on. The other members won't like it if you ask questions that have already been answered.

You can (and should) learn what a newsgroup is about by looking at its FAQ file—which stands for Frequently Asked Questions. FAQ files are really the *answer*s to those Frequently Asked Questions. FAQ files are often posted by newsgroups that are moderated, and they contain valuable background information about the newsgroup's topic. FAQs are designed to bring new participants up to speed without having to rehash everything that's been said before.

HOW NEWSGROUPS ARE ORGANIZED

You'll notice that newsgroups are organized—sort of—into major categories. Here are some of the categories that contain newsgroups on farming-related subjects:

alt.   "Alternative," the largest category of newsgroups focusing on popular culture, music, media, and controversial issues.

bit.   "Bitnet" newsgroups. Originally, these were mailing lists that gave birth to the UseNet newsgroup structure. Newsgroups that end in "bit" can cover just about any subject.

biz.   "Business," established for posting business-related messages.

clari.   "Clarinet," a wide-ranging source of newsgroups.

comp.   "Computer" newsgroups focus on computer and software.

gov.   Federal, state, and local "governments" and agencies (like USDA).

misc.   "Miscellaneous" newsgroups, which include newsgroups related to consumers, business, and legal issues.

rec.   "Recreation" includes discussions of hobbies, games, musical interests, and other recreational topics.

sci.   "Science" includes newsgroups focusing on science and technology.

Here are a few newsgroups that are ag-related. Just remember that all Internet Service Providers can choose which newsgroups that they carry. Your service may not carry these same newsgroups, but may carry similar ones.

    news:alt.agriculture.fruit
    news:alt.agriculture.misc
    news:clari.biz.industry.agriculture
    news:clari.web.biz.market.commmodities.agricultural
    news:gov.us.topic.agri.farms
    news:sci.agriculture.beekeeping
    news:sci.agriculture.ratites

SUBSCRIBING TO A NEWSGROUP

As we've said, before subscribing to a newsgroup, it's a good idea to go in and read some of the previous discussions to see if you really want to join. You should also read the FAQ file (if there is one) for the newsgroup because FAQs typically have valuable information for new users. In fact, there is an entire Web site for FAQs about newsgroups. The address is: **http://www.lib.ox.ac.uk/internet/news/faq/by_group.index.html**.

Moderated newsgroups may have strict rules for participation. Usually these rules reflect the general rules of Netiquette (see page 66), with a particular spin added by the newsgroup's moderator. One rule that almost all newsgroups abide by is to never post messages to the entire newsgroup that are intended for just one person. Generally, when you post an article to a newsgroup, your online response goes to the entire newsgroup. However, if you want to respond to a particular individual, you should send a private e-mail message. And it's even worse to send a "test" message to a newsgroup just to make sure "everything is working." In fact, there are two newsgroups that have been set up just for this purpose. If you want to test your system and insure that your postings are really going out over the Net, send them to alt.test or misc.test, then check either of these test newsgroups to find your test message.

SUBSCRIBING TO A NEWSGROUP USING NETSCAPE (3.01)

After you've told Netscape the name of the news server you'll be using, you will need to specify which newsgroups you want to read. There are several ways to do this. The easiest way is to use the list on page 146 to identify farming-related newsgroups of interest to you.

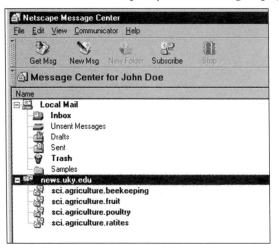

Newsgroups as displayed by Netscape.

The following instructions should work for both Windows and Macintosh versions of Netscape. First, pull down the **Window** menu and select **Netscape News.** Next, pull down the **File** menu and select **Add newsgroup.** Now, simply type in the name of the newsgroup you want to read.

If you don't know the name of a newsgroup (or simply want to see what's available on your ISP's server), you can have Netscape download a list of all of the newsgroups maintained on your ISP's news server.

First, go online and from Netscape's Web browser, pull down the **Window** menu and select **Netscape News**. When the Netscape News window opens, there may already be a few default newsgroups visible. But you want the whole list, so pull down the **Options** bar and select **Show All newsgroups**. (If there is more than one news server available, you may have to first click on the news server you want to use so that Netscape can identify it.)

Now, go make a cup of coffee while Netscape downloads the names of the several thousand newsgroups stored on your ISP's server. When that's done, you can scan the list and select any newsgroups that you want to read. Note that the names of lists with an asterisk are "nested" meaning that there are many more newsgroups under that category name. You can easily identify nested newsgroups because they have a little triangle pointing to them. (To open a nested newsgroup, click on the little triangle until it points down.)

Once you've found a newsgroup you want to read, place a check mark next to its name. When you click on the name of a "checked" newsgroup, all of the messages that have been posted to it will be downloaded. You'll see a counter that tells how many messages there are and how many you haven't read. If this is your first time, all of them will be "unread."

To read a message, simply click on it. To post a message of your own, click on the **To: News** button (to send a private e-mail message, click on **To: Mail**). After you're through reading, you can click on the **Group** button which will mark all of the messages as having been read. The next time you read the newsgroup, you'll only get the new (unread) postings. If you have any trouble, use Netscape's help system. To quit or unselect a newsgroup, reverse the above procedure, and this time click on the check mark until its gone. That will deselect the newsgroup.

SUBSCRIBING TO A NEWSGROUP USING EXPLORER

After you've told Explorer the name of the news server you'll be using, you will need to specify the names of the newsgroups you want to read. As with Netscape, there are several ways to tell Explorer which newsgroups you want to read. (Unfortunately, there are also several versions of Internet Explorer, and they differ in how newsgroups are identified and read.) The following instructions should work for most Windows versions, but keep in mind that the names and locations of some of the buttons may be different on the version you have. If you get stuck, use Internet Explorer's built-in help system.

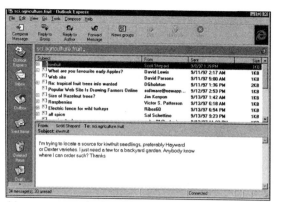

Newsgroups as displayed by Internet Explorer's *Outlook Express*.

First, go online and from Explorer's Web browser, click on the **News** button (or the **Mail** button, depending on which version you have). Next, click on **newsgroups**. Now, go make a cup of coffee while Explorer downloads the names of some 30,000 newsgroups (or however many are kept on your ISP's news server). When that's done, use the list on page 146 to identify the farming-related newsgroups on the list.

To select a newsgroup, simply type the name of the newsgroup in the "Display newsgroups containing:" box. As you begin typing, all of the newsgroups that don't match are eliminated until an exact match is found. For example, say the newsgroup you want to read is called "sci.agriculture.poultry." As soon as you type "sci" all newsgroups that are not "sci" are tossed out—but there are still a lot of newsgroups in the "sci" category. As you type "sci.agriculture" all of the "sci" newsgroups that don't have ".agriculture" are thrown out, and continue until you finally reach the particular newsgroup you want. By the way, you can stop typing at any point in this process and select any newsgroup visible on the list. For example, you could stop at "sci.agriculture" and select any newsgroup from the "sci.agriculture" list. Also, keep in mind that not all news servers contain every newsgroup, so your ISP's server may not hold a particular newsgroup you want to read.

Once you've selected a newsgroup from the list, you can read it. First, click on (highlight) a newsgroup you want to read. Next, click on **Go to newsgroup** and the current messages for that newsgroup will be loaded. To read a message, simply click on it. Most versions will automatically mark any messages you read so you'll know what you've read and haven't read. To manually mark all messages as having been read, pull down the **Edit** bar and choose **Mark All as Read** which will mark all of them as if you've read them.

To "subscribe" to a particular newsgroup (so you won't have to go through this selection process again and again), highlight the name of a newsgroup in the newsgroup window and click on **Subscribe**. If you have any trouble, use Explorer's Mail and News help system.

Once you've selected a newsgroup and have added it to your "Favorites" list, you read it by clicking on the **Mail** icon and selecting **Read News**. Next, pull down the **newsgroups** menu bar and highlight the one you want to read.

You can also read newsgroups by clicking on the **News** icon (from Internet Explorer) and pulling down the **News** menu and selecting **Go to Favorite newsgroup**. However, in some versions of Explorer, the **News** icon starts the newsgroup selection process (as described above), while the **Mail** icon merely starts the newsreader. As with the Mail system, there are buttons for handling tasks such as sending messages to the newsgroup (or to a particular message's author).

Macintosh versions of Internet Explorer work a little differently. To activate the newsreader system, Macintosh users will click on the little newspaper icon. If you don't already see a long list of newsgroups, pull down the **View** menu and select **Refresh**. Now, you will get a list of *categories* of newsgroups available on your ISP's news server. Click on a category to open it. Now, select a newsgroup you want to read by clicking on it. If you find a newsgroup you like and want to go to it quickly, pull down the **Favorites** menu and select **Add Page to Favorites**. The next time you want to read this newsgroup, pull down the **Favorites** menu and select **Go to Favorites**. Some Internet Explorer versions have icons on the tool bar that let you perform these functions without having to use the pull-down menus.

By the way, Explorer will let you quickly change news servers by clicking on the down arrow at the end of the news server box to select either your ISP's news server (which you set earlier in Preferences or by using Internet Config) or Microsoft's own news server (the default).

SUBSCRIBING TO A NEWSGROUP AT AN ONLINE SERVICE

Online services usually provide their own newsreader software, which is similar to Netscape or Explorer, but there are a few differences. First, the software from most online services doesn't need to be set up; that's usually done automatically when you install it. About all you have to do is go to the newsgroups area after you sign on, select a newsgroup, and read the messages.

To find a newsgroup on America Online:
From the Main Menu, click on the **Internet Connection** button. When the Internet Connection screen loads, click the **newsgroups** folder in the Resources list. Next click on **newsgroups**. From the newsgroup screen, click (again) on the **newsgroups** folder. (You can avoid all this clicking by using newsgroups as a Keyword from the GO menu.)

▼ *Chapter 5*

64

If you know the name of a newsgroup you want to subscribe to (say from the list on page 146), click on **Add newsgroups**. This displays a long list of newsgroup categories. Select the appropriate category. (For example, a newsgroup that begins with "sci" is in the **sci.** or science category.) Double-click any category to see newsgroups under that category. Double-click on a newsgroup to select it. Now you will get a list of messages by subject. Scroll down the list until you find a subject you want to read about and (finally) click on any messages you want to read.

If you don't know the name of a particular newsgroup, you can search for it by subject by clicking on **Search all newsgroups**. You'll get a box where you can enter the names of topics to search for. Try searching for "beekeeping" and see what you get. You can select one of these newsgroups and read it as described above.

Typical farming-related newsgroups from America Online.

If you find a particular newsgroup you like, you can add it to your list of favorites by clicking on **Add newsgroups**. Then, the next time you want to read your newsgroups, simply click on the **Read My newsgroups** button and you'll go to your newsgroups without having to search for them again. America Online and other online services have other user-friendly features like Expert Add that make finding and reading newsgroups very easy.

If you are on the Web (using your Web browser) and you know the address of a newsgroup you'd like to read, you can enter the URL for that newsgroup just as you would for any other Web site address, except that the address begins with news: (instead of http://).

NETIQUETTE

"Netiquette" is short for Net etiquette. You will find that there are (mostly) unwritten rules for communicating with others on the Internet. These guidelines are based on common sense and common courtesy. They are good rules of thumb for any form of verbal communication—online or off—but they are particularly appropriate to Internet communication.

Here are some of the basics of Netiquette:

Typing something in all caps is the equivalent of SHOUTING.

"Flaming" is the rather unattractive practice of responding to a message with an inflammatory, rude, and insulting reply. It's not good netiquette to flame, so type unto others what you would have them type unto you.

Few e-mail programs have spell checkers, so it is considered improper to comment on someone's typing or spelling. It's just electrons, after all.

There are several cute symbols you may see in e-mail messages that look like bad punctuation. Here's the symbol for a smiley face :-) (You'll have to look at it sideways.)

Some e-mail messages from long-time Internet users contain many abbreviations or emoticons with which you may not be familiar, such as:

ATM	**At the moment**
BTW	**By the way**
F2F	**Face to face**
FAQ	**Frequently Asked Question**
FYI	**For your information**
IMHO	**In my humble opinion (used in newsgroups to avoid getting flamed when expressing a strongly held view)**
LOL	**Laughing out loud**
R	**Are**
ROFL	**Rolling on the floor laughing**
TIA	**Thanks in advance**
:-)	**Happy Face**
;-)	**Wink**
:-(	**Sad Face**

CHAPTER 6
CHAT ROOMS

As we discussed in Chapter 3, e-mail works very much like "snail mail" in that someone sends you a letter that goes into your "mail box" and sits there until you read it. But this can be one of e-mail's downsides, too—it is very difficult to conduct a "conversation" via e-mail. To converse in "real time" with someone on the Internet, you need a system that works more like the telephone than the mail box. One way to have a conversation with other people on the Internet is to visit a chat room. (While it is possible to actually speak to other people on the Net, this requires special "Internet telephone" software.) To "talk" to someone in a chat room, you use the keyboard and type what you want to say—in other words, you let "your fingers do the talking."

To talk to someone using the phone, both of you must be "online" at the same time. This is also true with chat rooms. All of the participants are online at the same time.

In a particular chat room, usually only one subject is being discussed. The subject could be very broad such as "precision farming" or it could be very narrow like "GPS systems." Some chat rooms, like the *Farm Journal Today* site, host discussions on a wide variety of topics, while others hold a continuing discussion about just one thing, like antique tractors.

CHAT ROOM SOFTWARE

When the idea of online, real-time conversations first started on the Internet, about the only way to do it was by using a system called Internet Relay Chat or IRC. IRC is still used to this day, but to use it, your computer needs to have a special software program called an IRC client. One of the most popular versions of an IRC client for Windows is mIRC. However, Internet Explorer (currently) comes with an IRC client (of sorts) called Comic Chat. It's not great, but it's free. Another free IRC client that is pretty good is Netscape Chat. If you use Netscape as your browser, it may or may not already be installed on your computer, depending on which version you have installed. If Netscape Chat didn't come with your copy of the Netscape browser, you'll need to get a copy of the software by going to Netscape's Web site (**home.netscape.com**) and downloading a free copy of Netscape Chat. Also, as we've said before, by the time you read this, Internet Explorer may have a newer and better IRC client (or none at all) and ditto for Netscape. However, as has also been mentioned many times before, many if not most Internet applications are going away from using stand-alone software and special protocols and are being moved to the Web (where all you need is a Web browser—see Chapter 7). If the chat room has a Web address (see below), all you may need is a Web browser in order to chat.

FINDING CHAT ROOMS

To use most other Internet services, you only need to know the place—an e-mail or Web address. To use chat rooms, you not only need to know where they are (the address on the Net where you can find them), you also need to know when they are being used. Some chat rooms are open and available 24 hours a day, although they may be empty during regular business hours or very late at night (depending on the subject being discussed). Most chat rooms are open for business at a particular time and some are open only on certain days of the week. For example, Farm Journal sponsors a popular chat room called Monday Night Campfire where all kinds of farming topics are discussed. It's "open" every Monday night at 8 p.m. (ET). To join in, simply go to **www.farmjournal.com**.

Although many chat rooms are available to anyone with access to the Web, some chat rooms operated by online services (like AOL) are usually for "members only." In other words, you would have to subscribe to America Online in order to join in a chat room discussion that's held on AOL—likewise with CompuServe. Occasionally, some online service chat rooms are also accessible to anyone on the Web. Obviously, if you use an online service as your Internet provider, you'll have access to that online service's chat rooms as well as to any on the Web itself.

USING CHAT ROOMS

Participating in live online discussions can be a little strange until you get used to how they work. It's important to realize that it takes a little time for your message to get posted (so other people can see it). It also takes time for others to read what you wrote and to type a response. It may take several minutes. Be patient. A short delay doesn't mean that no one is interested in what you just said. Some people read and type faster than others.

Since it takes awhile to read and respond to messages, don't try to be too "conversational"—the lag time is much too long to try to conduct back and forth dialogue. Also, forget those rules of etiquette your parents taught you, like "keep it short" and "don't interrupt." These rules were meant for face-to-face conversations, not the Internet. You don't have any control over when your message gets posted, so when it finally does show up, you're not really "butting in" to someone's conversation.

When you type a question or a response, explain yourself fully and include all pertinent information. While you don't want to appear to be "long winded," you need to put enough information in your message so that readers can understand your question or comments. For example:

> I have about 40 acres of mature oak and walnut timber on an easily accessible farm. What price should I expect to receive when I sell the timber, and what problems should I watch out for?

Another thing that will take some getting used to is having several conversations or *threads* going on at the same time. You might see a message on one topic, and while you are typing a response, several messages about different subjects will appear "out of order." This is normal. There usually will be several, sometimes unrelated, threads scrolling up the screen at the same time. Disregard the subjects and messages that don't interest you. It may seem like a jumble at first, but you'll get used to it. Lastly, don't fret over typos and spelling—all of this is being done as they say, "on the fly."

Chat room conversations make extensive use of abbreviations and *emoticons*. The latter are symbols such as the Happy Face :-) which are made with the creative use of punctuation marks. For a list of the more commonly used abbreviations and emoticons, see page 66.

Chat rooms can be fun, but occasionally they can be dangerous. Every now and then, the media run stories of someone who has been defrauded, kidnapped (or worse) by someone they "met" in an Internet chat room. If you met a stranger on the street, you probably wouldn't feel comfortable giving them your real name and address, your phone number or your credit card numbers, passwords or other personal information. You probably wouldn't agree to meet them at night in another part of town either. These are good practices to follow in chat rooms, too. Given the vast number of chat room

conversations that are going on at any given moment, chat rooms are probably statistically safer than the parking lots of most convenience stores. But on the Internet, you never know to whom you are talking. A guy you meet in an Internet chat room who says he's the Pope could really be His Holiness but he could also be—and in this example, probably is—a con artist. Don't take chances. Use only your first name (or one you made up), and never give out personal information. After many months of using a particular chat room, you may feel you know everyone there. Maybe you do know them. Maybe they are all right. But since someone can visit a chat room without saying anything or otherwise making their presence known, you can't be sure about who else is in the room. If you need to share personal information with someone you really do know, send them a private e-mail message instead of blurting it out in the chat room—even if (you think) there are just the two of you in the room.

Tom (nc IN)	Does irradiation create a "sterile" product? Is that what we're really shooting for?	10/9/97 8:07:31 PM
Robert LaBudde	Actually, there's not much that can be done quickly. The big packers are already converting to slaughter interventions strategies, and HACCP is starting in Jan. 1998.	10/9/97 8:08:18 PM
Steve Suther	Will we irradiate all produce as well as meat in a year or so? Just take us one at a time, Robert. ;-)	10/9/97 8:08:26 PM
Robert LaBudde	Tom: irradiation for meat is only going to be pasteurization (partial kill), not sterilization for normal food.	10/9/97 8:08:56 PM
Steve Suther	By slaughter intervention, do you mean screening of high-risk cattle, or what?	10/9/97 8:09:41 PM
Robert LaBudde	I wouldn't hold my breath waiting for produce to be irradiated, although high-value, high-risk foods (such as raspberries) might be a possibility.	10/9/97 8:09:49 PM
Robert LaBudde	Steve: No, clean-up procedures in the packing plant, such as steam pasteurization of dressed carcasses or washing with organic acids. These clean up contamination after it's happened.	10/9/97 8:10:58 PM
Steve Suther	Any chance we will have a device about like the box over the conveyor belt at the airport so food goes through and gets zapped? Or what form will it take. . .?	10/9/97 8:11:42 PM
Steve Suther	I read that cobalt 60 and cesium 137 are the two main elements considered for irradiation. I can see the standup comedian saying we are more concerned about half-life than shelf life now. But seriously, will consumers go for irradiated beef?	10/9/97 8:14:10 PM
Robert LaBudde	Currently there are only a few reasonable capacity general purpose irradiators in the US. They couldn't handle any significant new volume.	10/9/97 8:14:27 PM

A typical live "chat" session from *Farm Journal's* Web site.

SAMPLE CHAT ROOM SESSION

Here's what a typical session in a chat room looks like. For our example, we'll use *Farm Journal Today's* chat room. Here, on Monday nights at 8 p.m. (ET) you'll find farmers talking with other farmers about a variety of agricultural issues. To get to this chat room, we'll first go to the home page of *Farm Journal's* Web site (see Chapter 12). If you already have your Web browser running, you can follow along. The Web address is **www.farmjournal.com**. After the home page loads, there are several options available. One of the options is **Live Chat**. Click on that button and you'll be asked for your name and e-mail address. Use your first name and your state (or a nick name). You may also be asked which chat room you want to join. Sometimes there are several different discussions all going on at once. Most discussions are held in the General Auditorium.

In the screen shot on the bottom of the opposite page, you can see (on the right side of the window) how many people are already in the room. On the left side of the window, you can read what's been said during the past several minutes.

At the bottom of the screen, you have a little box in which you can type what you want to say. After you have typed your message, you must click on the **Post** button to send your message to the chat room. It will usually be posted for all to see in a few seconds. If the room is very busy, this could take a while. About every 30 seconds or so, the screen will update (refresh) and you'll be able to read everything that's been posted since the last update. If you can't wait for the automatic refresh, you can always click on the **Refresh** button and get an instant update.

Other chat rooms on the Web work pretty much the same way. You go to an address, select a room, read what's been said, write and post messages and (if necessary) update your screen from time to time.

CHAPTER 7
THE WORLD WIDE WEB

Without a doubt, the most popular (and most fun) way to access information on the Internet is the World Wide Web. CERN, the European Physics Laboratory in Geneva, Switzerland, created the World Wide Web in 1989 as an easier way of making data available to people all over the world. They wanted users to be able to access information of any type from any source in a simple and consistent way. As is true of the Internet in general, there is no central location where the Web resides—it's all out there somewhere in *cyberspace*. The Web is composed of computers scattered all over the world, linked into networks mainly via high capacity digital lines, to form networks of networks.

The Web offers access to information virtually anywhere on the Internet and in virtually any format. Although the primary format for information on the Web is HTTP (which stands for HyperText Transmission Protocol), you can also use the Web to access other file types including FTP (File Transfer Protocol) and even access newsgroups, too. The Web also has audio files you can listen to and even video clips you can view.

The primary difference between the Web and the rest of the Internet is in the presentation. The other services we discussed in earlier chapters are largely text-based. The Web uses a graphical interface called a *browser* which allows you to navigate (browse) the Web using point and click techniques (like Windows or the Mac) instead of having to know (and type in) a lot of commands.

Information on the Web is typically displayed in pages written in a special format called HyperText Markup Language (HTML). You'll see a lot of Web addresses that end in html, htm or some other variation. You don't need to know anything about HTML in order to use the Web. In fact, it's HTML that makes using the Web so easy. A page written in HTML can (and usually does) contain highlighted words that are connected (linked) to other pages on the Web. So to move from one page to another, you simply click on highlighted words or icons—there's surprisingly little typing to do. In that way, the Web is a lot like Windows (or the Macintosh), while the rest of the Net is more like plain old text-based DOS.

WEB SOFTWARE

The software you'll need for the Web is called a Web browser. As discussed in earlier chapters, the newest versions of most Web browsers like Netscape and Internet Explorer also handle other Internet services like e-mail, newsgroups, etc.

You can probably obtain a Netscape or Internet Explorer browser for free—your ISP will probably provide one or the other as part of a "start-up package." If not, both browsers can be downloaded from the Web at no charge. However, this presents a classic "chicken/egg" problem since you obviously must have access to the Web to download a browser, and a browser (or FTP program) is needed to download anything from the Web. Another complication is that although both browsers can be downloaded at no cost, Internet Explorer (at this writing) is totally free while Netscape (at this writing) requires that you send in a small fee if you want to use their browser. This payment method works on the honor system. If you have access to a friend's computer (who is already on the Internet), you can download a browser. The Web address to download Internet Explorer is **www.microsoft.com** and Netscape's address is **www.netscape.com**.

The easiest way to obtain a browser is to buy one of the "official" guidebooks from either Netscape or Microsoft. These books include complete, full-power versions of each company's Web browser (usually on CD-ROM) and may include additional software "plug ins" that add functionality to the Web browser (allowing you to hear live audio over the Net, for example). In addition to having a complete version of the browser's software, you'll also get a complete operating manual which will help you install and use the browser like a pro. Most bookstores stock these books or can order them for you, but be sure to get the version for the computer/operating system you use (Windows 3.1, Windows 95, Windows 98, Macintosh, etc.). Some books, like Microsoft's *Official Internet Explorer Book*, cover Windows 3.1, Windows 95 and Macintosh, while others, like the *Official Netscape Navigator Book*, come in various "flavors" depending on which operation system and computer you use.

In any case, <u>check with your intended ISP</u> before buying a Web browser. They may provide one as part of a start-up package or have some unusual operating constraints that make one browser (or the other) easier to use.

▼ *Chapter 7*

If you use an online service, they may provide their own Web browser (which looks and works a lot like Netscape and Internet Explorer), or they may actually use special versions of Netscape or Internet Explorer as their "official" Web browser. If the software for your online service is on CD-ROM, it probably includes a browser, but if it's on a floppy disk, you may have to log onto the online service and download the browser (because it's too large to fit on a single floppy). Since this can take quite a while (a half an hour or longer), it's best to obtain the CD-ROM version of the online service's software if your computer has a CD-ROM drive.

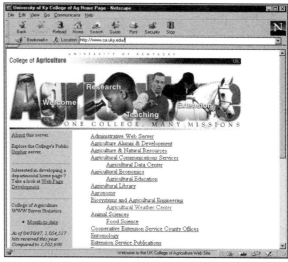

The University of Kentucky, like most land grant universities, goes to great lengths to provide farmers with timely and accurate information on a variety of subjects.

After you've used a browser for six months or so, an updated version has probably been released with more features (and a few of the buttons moved around). You can update (get the latest version of) your Web browser simply by clicking on the browser's icon. In Netscape, it's the big "N" at the top of the window, and in Explorer, it's the big "E."

HELP!

While writing this book, we used the phone, the mail, FedEx and UPS quite a lot. They all performed well almost all the time. But, as you know yourself, no system works flawlessly every time. Sometimes when you make a call, you get a wrong number. Frequently, it's because you misdialed the number, but occasionally the phone system itself connected you to the wrong number even though you dialed the "right" one. Sometimes there's noise on the line. On rare occasions, you pick up the phone and there's no dial tone.

The Internet, like all other complex systems, does not work flawlessly all the time. If you are new to the Internet, these operational errors can be confusing. You might type in an address of a Web site and the browser might respond with a message that says "The domain name server does not exist or could not be found." Or the message could state that the page you're looking for doesn't exist or has been moved to another address. The fact that a page couldn't be found (or doesn't exist) could be because you typed the address incorrectly, or it could be because of some transmission fault. When things don't work, it's hard to know if it's your fault or the Net's fault.

If you are having trouble getting your browser to work properly, you should make use of the help systems built into both Netscape and Internet Explorer. However, the help files may not have been installed on your computer when you installed the browser and may need to be downloaded from the Web. This can be a problem if you can't get your Web browser to work at all. If that's the case, give your ISP's customer service department a call.

A Web browser that doesn't seem to want to work at all is probably due to the inability of your computer to make a proper connection with your ISP. This is likely due to a problem with the TCP/IP or dialer program, for which you may need help from your ISP to fix. If your browser does work (meaning you can at least connect to the Web and download something) but it doesn't work like you think it should or you can't figure out how to use a certain feature, then the browser's help system is there to help you.

One final (and oft repeated) warning concerning the specific examples that we use in this book is that the particular version of either Netscape or Internet Explorer you are using may be different. These differences will probably be insignificant and may amount to nothing more than a button being in a different place or a menu selection having a slightly different name from the examples we use. But, occasionally, a browser feature may vary quite a bit from our example. Therefore, don't be surprised if your version of Netscape or Internet Explorer has a somewhat different layout or operates a little differently from the way we've described it. Even the "official" guides can't keep up with all the changes. On the other hand, the built-in help systems do seem to be up-to-date, so use them if you can't find a particular button or feature we've described.

HYPERTEXT

For most people, the World Wide Web *is* the Internet. The Web links information resources using a point-and-click interface that lets you click on words or pictures (icons) and go directly to those resources. You don't need to know any special Internet commands or how the system actually works.

The thing that really drives the Web is a programming language called HyperText Markup Language (HTML). You don't need to know how to create pages using HTML just to use the Web. All you need to know about HTML is that it allows the creator of the page to create active or HyperText links (which are usually displayed in a different color from the rest of the text). If you click on a highlighted HyperText link, the browser will activate that link and go to another page somewhere else on the Web. Sometimes HyperText links can be used for other things, such as taking you to a form so that you can enter your name, e-mail address, a subject to search for, or other information.

Clicking on the highlighted (or underlined) words will take you to another page that could be anywhere on the Web.

If you are impatient, you can click on a HyperText link (and go somewhere else) as soon as the link is visible on the screen, without waiting for the entire page to load.

CREATING YOUR OWN HOME PAGE

To create your own Web home page, you must use the Web authoring language called HTML (HyperText Markup Language). Technically speaking, the first page is called the home page, and the entire collection of materials stored on a server is called a Web site.

Creating a home page is not expensive or difficult. In fact, some of the newer word processors will help you do that. (You will, however, need to place your home page on an ISP or online service's server in order for other people to have access to it. Check with your ISP or online service for more information.)

WEB ADDRESSES

Information on the Web is usually kept in documents called pages. These pages (or files) are stored on a computer somewhere in cyberspace. (The computer is called a Web server while the collection of Web pages on the server is called a Web site). There are literally hundreds of millions of pages on the Web. To find a particular page out of millions, you'll need to know the address of the page you want to see (or Web site you want to visit). On the Web, an address is called a URL which stands for "Uniform Resource Locator." Most of this book is devoted to listing farming-related addressees (URLs) beginning on page 119. Or you can search for them on your own using the search engines described in Chapter 11.

There are several ways of telling the Web browser the address or URL of the page you want to obtain. As discussed above, your Web browser already knows at least one address—its default home page. To go to other places on the Web, you'll need to enter other URLs. You can open a **Go to Location** (or **Open Location**) box from a pull-down menu and type in a URL or you can click on a highlighted HyperText link. You can even type over the address that's in the current location box and press return (or enter).

Obviously, clicking on a HyperText link is easier and faster than typing a URL. When you click on a link, the URL for that link is displayed in a little box on your Web browser. In fact, most of the time, you don't need to know the URL to get to the page you want—you just click a highlighted link on one Web page to go to another and another. You can go from link to link on the Web without ever having to know or type a URL.

The next time you're using your browser, place the pointer over a link and look for the browser to display the URL (usually at the bottom of the screen). The URL will look something like this: **http://www.rural.org**.

Most browsers have a "Go To" box where you can type in a Web address, press return (or enter), and there you are. Usually, you don't have to type in http://. Most browsers will add that automatically.

HOW URLS WORK

The first component of a URL is the protocol that identifies how the information will be transmitted over the Net. On the Web, most pages are written in HTML, so the protocol is usually HyperText Transmission Protocol. Therefore, the first part of most Web addresses is **http://** (the colon and forward slashes are important). As discussed earlier, some Internet addresses begin with **ftp://** (for files to be transmitted in File Transfer Protocol), **news:** (for accessing newsgroups over the Web), or **gopher://** (for accessing gopher sites). Gopher sites, by the way, are menu-based Internet information libraries.

The second component of a typical URL identifies the server. In this example, the **www.uky.edu** identifies the page as residing on a Web server at an educational institution and the "uky" stands for the University of Kentucky. Most Web addresses start with the letters **www** in the URL, but some don't even have **www** anywhere in the URL and still point to pages on the Web. An example of a URL that doesn't include the letters **www** is **http://leviathan.tamu.edu** which is a Web site located at Texas A&M University.

Sometimes the URL for a site is extremely long and includes more information than the domain name. An example of this is **http://gnv.ifas.ufl.edu/www/agator/htm/ag.htm** where the last part of a URL is the pathname, and it identifies the location of the item on the Web server and the layers of directories (folders) that the browser must go through to get to the site. Each segment of the pathname is usually preceded by a single forward slash. The very last part "ag.htm" is the filename of the actual Web page that will be displayed on your screen. Page names usually, but not always, end with the letters html, htm, or shtml. Note the occasional use of odd characters, such as the tilde (~). Spaces are usually represented by the underline (_) character.

Thus, a complete Web address might look something like this: **http://gnv.ifas.ufl.edu/www/agator/htm/ag.htm**. Some Web addresses are much shorter and simpler like this: **http://www.cbot.com**.

ENTERING URLS

With most browsers, there are several ways to enter a URL. The fastest way is to move the mouse pointer to the box where your browser displays the current URL and type over the current URL. In Netscape this box is called **Go To**, while in Explorer it's called **Address**. In either case, you can't miss it because it is the long box at the top of the screen that has a URL already in it.

The second way is to use the pull-down menus. In both Netscape and Explorer, pull down the **File** menu and select **Open Location**. You'll get a new box in which to type the URL. Type it *exactly* as it is printed in this book. URL's are usually "case sensitive" which means that capital letters should be capitalized and lower-case letters should be lower-case. Also, don't confuse the zero with the letter "O" or the symbol for the number one with the letter "I."

Once the URL is entered, the browser will go out to the Net and try to find that page. You can see this in progress by looking at the large N in Netscape (or the large E in Explorer). If data is coming in, you'll see a "meteor shower" in Netscape or a revolving E in Explorer. As a page begins to load, you will see a display at the bottom right of the window that tells you how large the page is, how fast it's coming in, and approximately how long it will take to download. Since information on the Internet is sent out in little packets, the estimate about how long the download will take may not be very accurate.

Occasionally, you'll type in a URL (or click on a link) and get a message that says: "Server does not have a DNS entry. Check the server name (URL) and try again." This message could mean that you typed the URL incorrectly or that the URL doesn't exist. But it could also mean that the server you're trying to connect to is overworked and too busy to serve you promptly–try again later.

Many of the "error messages" you'll get on the Internet are vague. For example, your browser has no way of really knowing if a server exists or not. When you type in a URL, your browser starts timing. If the browser doesn't get a response back in a few seconds, it assumes that the server doesn't exist and will tell you something like "Server does not have a DNS entry." It could simply be that the responding server is slow. Check the URL and try again.

Also, keep in mind that things change quickly on the Net, and an address that is here today could be gone tomorrow. Sometimes Web sites move and the URLs are forwarded and sometimes they are lost in cyberspace. We do our best to keep an up-to-date list of farm-related URLs. The Web address for updates is **www.rural.org**.

SAVING URLS

Once you find a Web site you might want to visit again, you should store its URL so you won't have to search for it or type it again and again. You can do this by adding the site to a list of favorite or frequently visited sites. In Netscape, this custom list of URLs is called "Bookmarks." Explorer calls them "Favorites."

To save a URL in Netscape, you must be viewing the page whose URL you want to keep. Next, pull down the **Bookmarks** menu and select **Add Bookmark**.

There are two ways to save a URL in Explorer. You can load the page and pull down the **Favorites** menu and select **Add Page to Favorites**, or you can pull down the **Favorites** menu and select **New Favorite** and type in a URL (without having to first load the page).

Once you've saved a URL, you can go back to that site by simply pulling down the **Bookmarks** (or **Favorites**) bar and selecting the name of the site you saved. The URL for that site will appear in the browser's **Go To** (or **Address**) box and the browser will go to that site and load the page requested.

USING YOUR WEB BROWSER

Both Netscape and Explorer are very similar in look and feel. Both have several buttons on the tool bar that will help you navigate the Web.

NETSCAPE

The Netscape tool bar has two rows of buttons. Starting at the top left, the first two buttons let you move back and forth between pages you've loaded. The **Home** button always takes you to "your" home page—either the home page Netscape comes with or one you've selected to be your home page (see page 84).

This is the main command console from Netscape. You can go "somewhere" by clicking on the Home icon, typing an address in the box, or pulling down the File menu and entering an address.

Sometimes pages don't completely load or there is a transmission error. In that case, you can click on **Reload**. To stop a download, simply click on the **Stop** sign. **Mail** and **News** are activated from the **Window** (or **Communicator**) pull-down menu.

INTERNET EXPLORER

Like Netscape, Explorer also has a tool bar with two rows of buttons. Starting at the top left, the first two buttons let you move back and forth between pages you've loaded. The **Home** button always takes you to "your" home page—either the home page Explorer comes with or one you've selected (see page 84).

The main command console from Internet Explorer is very similar to Netscape. You can go "somewhere" by clicking on the Home icon, typing an address in the box, or pulling down the File menu and entering an address.

Sometimes pages don't completely load or there is an error. In that case, you can click on **Refresh**. To stop a download, simply click on the **Stop** (X) symbol. **Mail** and **News** have their own buttons, too. (Some versions of Explorer use the **Mail** button to activate both mail and news.)

BASIC WEB BROWSER OPERATIONS

Here's a start-to-finish description of what happens when you use a browser to access the Web. By simply starting your Web browser (Netscape or Internet Explorer), you trigger a chain of events. First, the browser will command your TCP/IP and dialer software to call your ISP and initiate an Internet connection (assuming they have been installed and properly configured—see the appendices for more information). Once the connection is established, the browser will load a *home page*. The home page is the first page your browser goes to every time it's started. When you first install a browser, it has a default home page already defined. If you are using Netscape, the default home page is probably **http://home.netscape.com**. And of course, if you're using Microsoft's Internet Explorer, the default home page will probably be **http://home.microsoft.com**. Since the home page is where your browser goes every time it's started, you may want to change the default home page to some other (farming-related) location—see page 84.

Once the home page loads, you have four options: (1) you can click on a highlighted HyperText link, (2) you can type over the current URL, (3) you can pull down the **File** menu and select **Open Location** and type in a URL, or (4) you can open the **Bookmarks** or **Favorites** menu and select a URL you previously saved.

Everything from Hubble Space Telescope images to Letterman's Top Ten list is on the Web. Elementary schools, nonprofit organizations, rock groups, and even individual farmers have created their own home pages on the Web. These different pages of information on the Web are linked together with hyperlinks—computer commands that are programmed into the highlighted words or graphics you see on your screen so that you can jump to another page (which could be anywhere in the world) by just clicking on the linked word or icon.

When you visit most Web sites, you will often see links to other related sites that you can visit. These "hot" links (or HyperText links) are usually highlighted in a different color from the rest of the text or use graphical symbols or icons. To follow a link, just click the word or icon to go there. With most browsers, the link will change color to indicate you've already been there, which helps you keep track of what you are doing. The interconnectivity provided by hyperlinks, with multiple ways of accessing topics, makes it possible for the Web to provide all different sorts of information—data, graphics, audio, and video—in one easy-to-use interface.

CACHING AND GRAPHICS

Every time you enter an address (URL) your browser will go to the Web and load the requested page. Some of these pages contain graphics and other images that are quite large and may take some time to download. One way to speed things up is to turn the graphics off. If you are using Explorer, pull down the **View** (or **Edit**) menu and select **Options** (or **Preferences**). Now select **Web Content** and uncheck the **Show Pictures** box. (You can also uncheck the **Play Sounds** and **Show Video** boxes for even faster performance.) In Netscape 3.01, pull down the **Options** menu and select **Auto Load Images**. [NOTE: At this writing, Netscape Communicator won't let you turn off images.] The check mark will disappear, meaning that graphics won't be retrieved automatically. Instead, you'll get a little graphics symbol that has a circle, triangle and square. If you want to see what the graphic really shows, simply click on one of these graphics icons and that picture will be downloaded.

The little graphics icon (with a circle, triangle and square) means that there is a graphic there. To see the graphic, simply click on the graphics icon and the picture will load.

If you turn graphics off, your Web screens will load much faster, but of course, you won't see the graphics, which is the whole point of the Web. While most graphics are simply for show and don't need to be downloaded for you to get information from the page, some graphics (called image maps) must be displayed because you have to click on certain parts of the image in order to make a selection (to go to another page, for example). Even if you are using a browser in the "graphics off" mode, you can always load an individual graphic by clicking on the graphic icon (as shown above).

Another way to speed up the Web is by using a feature called *caching*. Normally, each time you move to a page, the page is reloaded (refreshed) from the Web. If you move back and forth between pages, you may have to wait while each page is loaded over and over. You can stop this from happening by telling the browser not to reload (refresh) pages you've already looked at (during the current online session). Instead, these pages will be stored and retrieved from your computer's memory, which is much faster than loading them again from the Web. In Netscape 3.01, pull down the **Options** window and select **Network Preferences**. In the **Check Documents** box, click the **Once Per Session** button.

(In Netscape Communicator, go to **Edit, Preferences**, double-click **Advanced**, then select **Cache**. Click on the **Once per Session** button.) Now, once a page has been retrieved from the Web, it will be stored in your computer so that the next time you want to see that same page (during this session) it will be loaded from your computer rather than from the Web. You may need to increase the amount of cache memory allocated to this feature if you plan to load and look at a lot of pages. The cache memory setting is above the **Check Documents** box. You can also click on the **Clear the Disk Cache Now** button to clear what's there if you need more cache room and don't have the disk space available. If you click on the button marked **Every Time**, then Netscape will reload the page from the Web every time you want that page. If you click on **Never**, then Netscape will never update the page. This can cause some confusion because if the page changes from day to day and you "load" it again sometime in the future, you'll get the old one you saved rather than today's version.

Explorer works much the same way. Pull down the **View** (or **Edit**) menu and select **Options** (or **Preferences**), then select **Advanced**. Now, click on the settings box for **Temporary Internet Files** and click on **Everytime You Start Internet Explorer** (sometimes also called **Once Per Session**). You can also use this menu to adjust the size of the cache or to empty it. Both Netscape and Explorer's online help systems contain other tips that can help you speed up your browser's performance.

THERE'S NO PLACE LIKE HOME

Each time you start your Web browser, it looks up the URL for the default home page, and then it retrieves that home page from the Web. To go to other places on the Web, you can click links embedded in the home page or you can type URLs.

The default home page (the one that comes with your Web browser) is usually linked to the software company that makes the browser (or to your Internet Service Provider's home page). This may not be very useful to you.

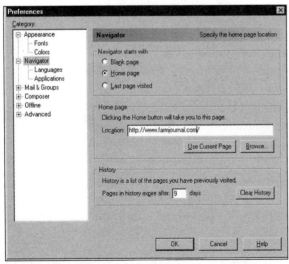

Farm Journal Today (**www.farmjournal.com**) is a favorite home page of thousands of farmers. You can change your browser's home page to make it start where you want to.

It's often better to go into the browser's preference file and change the home page to a place from which you would rather start—like **www.farmjournal.com**. To change the default home page in Netscape, pull down the **Options** menu and select **General Preferences**. The home page location box is marked "Browser Starts with." Type in the new address for the home page you want.

In Explorer, first go to the Web page that you want to use as your home page (such as **www.farmjournal.com**). Next, pull down the **View** (or **Edit**) menu and select **Options** (or **Preferences**). Then select **Navigation** and **Home/Search Page** and click on **Use Current**. Now, your default home page will change to the one that's currently in your Web browser's window.

SAFETY IN NUMBERS

The Internet is not very secure. Anything (like e-mail) that's sent over the Internet without first being encrypted can, in theory, be read by someone else. This doesn't happen that often, but the possibility should alert you to the potential problem of giving out your credit card numbers, bank account numbers and other financial or personal information over the Internet. Most credible online shopping services employ encryption systems of one kind or another.

Modern Web browsers like Netscape and Explorer have encryption systems built in. In Netscape, look at the little key at the bottom left-hand corner of the browser window. If the key is "broken," you don't have a secure connection (and you won't need one most of the time). When the key is solid (unbroken) you have a secure connection and the information you send will be encrypted. In Explorer, you'll see a "closed lock" symbol when you are in the encrypted mode and your communications are protected.

Occasionally, you will read about a group of college students who have broken some browser's encryption system. Virtually any encryption system can be broken if enough computational resources are applied. Nonetheless, the encryption systems built into browsers like Netscape and Explorer provide an acceptable level of security for conducting personal transactions (such as buying something with a credit card) because the cost of breaking the code exceeds the "value" of obtaining your credit card number by a wide margin. If someone obtains your credit card number and charges items to your account, most states limit your liability to $50. In any case, if someone really wants to steal your credit card number, it would be easier to go through the dumpster at the mall than to break Netscape or Explorer's encryption system.

ACCESSING FTP SITES FROM THE WEB

FTP stands for File Transfer Protocol, and it is a method of getting files from someone else's computer into your computer (or, sometimes, vice versa). You can use e-mail, mailing lists, newsgroups, and most of the World Wide Web to transfer files, but those are principally text or graphics files. Software and other files intended to be "read" by a computer are usually available via FTP. (However, just about any type of file could be accessed via FTP.)

FTP is a particular method of accessing files on a remote computer. FTP was created so people could copy files from one computer—also known as an FTP site—to another across the Internet. You can use FTP to get documents, software, text, graphics, or even sound and video clips.

Using FTP is a little different from using other Internet services. You actually log on to a remote computer and use that remote computer's resources to find and get files. Since you have to log on to the remote computer, you may need to have an account (or permission), but most FTP sites allow outsiders to log on using the word "anonymous" as the user name and your e-mail address as the password. This is known as an "anonymous file transfer."

To access an FTP site using your Web browser, simply type ftp:// (followed by the Site's address) instead of http:// as you normally would when going to a Web site.

The software for FTP was developed long before graphical interfaces (and Web browsers), and so it uses obscure commands. Fortunately, with most Web browsers and online service software you do not have to know FTP commands to download files from the Internet. This is good for users since using FTP is not as simple as getting e-mail or using a Web browser's point-and-click system. With FTP, you are taking something from another computer (as opposed to being given an e-mail message or a Web page). However, since you can use Web browsers to access most FTP files, it can be so easy to get an FTP file that you may not even be aware you are using FTP to get them. The online services integrate FTP functions into their software seamlessly, as well. If your Web browser

▼ *Chapter 7*

doesn't handle FTP, you can use a stand-alone FTP program like Fetch. Most Web browsers like Netscape and Explorer will only get files via FTP, not send them.

To FTP using your Web browser, simply type the name of the FTP site in the box where you would normally type the URL (using ftp:// instead of http://). It's not unusual for a home page on the Web to link to an FTP site. In those cases, you simply click on the link and you're there. Hence, it's possible to use FTP without knowing how—and without even knowing that you are FTP-ing.

Nonetheless, there are a few differences between FTP and the Web's main protocol, HTTP. First, FTP is a text-based method of moving files, so don't expect much in the way of graphics. Since FTP is a way to transmit files (as opposed to simply viewing them), FTP files can be in any format and may even be compressed. As a result, FTP is usually reserved for getting software and other kinds of files that are meant to be read by the computer rather than by you. When you use FTP, be sure to get the file in a format you can use (and if it's compressed, in a version you can expand). After the file has been transferred, it will end up on your computer's disk drive, not on the browser's screen.

ANONYMOUS FILE TRANSFERS

In the past, to use FTP you had to have an active account (permission) on the remote computer system where the file you want is located. In most cases, you still do, but now virtually all FTP sites accept anonymous log ins. For example, you can log on anonymously to one of USDA's computers and use FTP to download the information from that site. One address to try is: ftp://esusda.gov.

Getting Windows '95 to remember passwords (so you don't have to type them in every time) can be difficult. Although the Windows '95 dialer program will remember your user name, it won't normally remember your password even though you checked the "Remember My Password" box. The reason Windows '95 won't remember your password is because you first have to set up a password file for your computer in order to give the dialer program a place to store passwords.

To create this password file, go to **Control Panel, Passwords** and click **Change Windows Password**. Now, set a simple password for your computer like "123" and restart your computer. Your computer will ask for a password (123). Enter it and click **OK**. Next, connect to your Internet Service Provider, enter your Internet password and now check the Remember My Password box. Next, go to **Control Panel, Passwords** and click **Change Windows Password** again. This time, type the password that you just set (123) in the old password box and don't enter anything in the New Password or Confirm New Password boxes. Now, when you restart your computer it will not ask you for a password, but since you have created a password file, the Internet dialer will have a place to store your Internet password, so it will remember your password if you now click the Remember My Password box.

CHAPTER 8
AGRICULTURAL WEATHER

On a morning in late December, the crew of a Peruvian fishing boat hauls in its nets. The catch this morning is good; in fact, it's huge. Since it's nearly Christmas time, the fishermen call this "gift" from the sea El Niño—The Christ Child. The unusually large catch is due to an annual warm water current that flows near Peru around Christmas time. Today, meteorologists use the term El Niño to refer, not to the annual warm current, but to a larger shift of warm Pacific water that happens every three to seven years. (When a true El Niño happens, fishing off the coast of Peru actually gets bad.)

What does the size of the catch of fishermen off the coast of Peru have to do with you? Plenty, we are now finding out. El Niño means the ocean temperatures in the tropical Pacific are warmer than normal. Sometimes, the waters are colder than normal (this is called La Niña). The back and forth "seesaw" of ocean temperatures in the Pacific is known as the Southern Oscillator, and it may have more impact on American agriculture and world commodities prices than anyone dreamed. A new study published in *Climatic Change* suggests that El Niño events can change the yields of corn, wheat, soybeans and other U.S. crops by as much as 30%. In addition, El Niño can also cause droughts in Brazil and parts of Africa and Asia, affecting food production in those areas, which in turn can have a major impact on global markets. (And you thought you only had to worry about the local weather.)

The fact is, that probably nothing is more important to farming than the weather, yet, ironically, the National Weather Service got out of the agricultural weather business in 1996 due to "budget reductions." The Weather Service's new mission is "primarily for the protection of life and property," and the more specialized services, like agricultural weather, were farmed out to private companies. The good news is that farmers today have more access to better weather information than ever before. Many large farming operations already subscribe to private ag weather services, but you can get much of the same information over the Internet for free. There are, of course, differences between the weather services you pay a monthly fee for and the weather you can get free off the Net. The primary difference is that private ag weather services can automatically alert you to changing weather conditions that affect the particular crops you produce, and their information is specifically tailored for farming. While some of the weather information available from the Internet is specific to farming, the Net won't alert you—you'll have to identify those weather sites that have ag weather (or weather reports you can use). Then you'll have to check those sites yourself—probably every day (which is why you should take some time to identify weather sites you can use and then "bookmark" those sites so that you can get to them easily and quickly).

AGRICULTURAL WEATHER ON THE INTERNET

If you know where to look, access to the Internet will get you Doppler weather radar, satellite images, soil temperatures, rainfall amounts, drying conditions, long-range forecasts and all kinds of other weather information updated frequently, or sometimes even "live." But finding really useful farming-related weather information quickly from the Internet can be difficult and frustrating. While it's relatively easy to use the Internet to get current conditions and three-day forecasts for major cities or reports of airport delays, finding sources of weather information specifically tailored for farming is not so simple. It may require spending several hours and visiting many Internet weather sites before you stumble on weather information you can use to help manage your farming operation.

There are more than 266 million Americans and less than *one percent* of them farm. About as many people fly on any given day as farm all year, so most weather information is designed for the people who live in cities and are concerned about storms, air travel, commuting, skiing, fishing or boating.

Sometimes, a quick glance at the Weather Channel is all you need, but to make business decisions about planting, harvesting, or pesticide application, you'll need more precise information than you can get from the weather map on the back page of *USA Today*.

What you really want is agricultural weather that covers your farm. Better yet, the weather information should directly relate to the type of operation you have. The wind chill index may not be important if you grow corn, but it is if you're running cattle. Likewise, the number of degree days for the European Corn Borer isn't that useful to ranchers (unless they also grow some of their own feed), but it is obviously very important to corn growers. Unfortunately, there is no Internet address called "weather@myfarm," but there are good sources of agricultural weather on the Internet, if you know where to look.

A good place to start from is, of course, *Farm Journal Today's* home page **www.farmjournal.com** (see Chapter 12). Or you can use our list of Weather Web addresses (see page 278) that we regularly update at **www.rural.org**. Depending on the particular weather site, you can get current conditions, three- and seven-day forecasts, Doppler weather radar, drying conditions, rainfall totals and other information.

ALL WEATHER IS LOCAL

While there are several very good weather sites on the Internet, you can run into problems trying to access some of the more popular national weather sites. For example, say you farm in Idaho and you've bookmarked a national weather site such as Intellicast (**www.intellicast.com**). When there is a hurricane off Miami, Intellicast's Web server is going to be accessed (or "hit") by thousands of people in south Florida who are tracking the hurricane. Computers can handle only so many hits at one time, so you may have trouble connecting during peak times, and when you do connect, downloading a weather map may be slow because so many people are trying to get the same thing from the same computer. This is not to say that you shouldn't use national weather sites, but it is a reminder that like politics, all weather is local. Often, the "best" weather sites are the ones located nearest to you. As in the example above, an Idaho weather site is not likely to be swamped due to a hurricane off Florida or a nor'easter off New England. Unfortunately, local weather sources are not as easy to locate as the larger, national weather services.

Current weather conditions and forecasts for cities large and small are fairly easy to find on the Internet (see Chapter 11). However, many national weather services require you to know not the name of the nearest city but the three letter code for the airport where the local weather service office is located. Thus, the code for Nashville, Tennessee is BNA, while the code for Lincoln, Nebraska, is LNK. If you're not sure of the code for the airport closest to you, look on your luggage for old airline tags or check with the closest National Weather Service (NWS) office (see page 284). In some cases you can simply click on the national weather maps to get information about specific cities without having to know the airport's code letters.

THREE KEY WEATHER SOURCES

There are three key local weather sites you should locate and bookmark. The first is from the NWS itself (**www.nws.noaa.gov**). While it's officially out of the ag weather business, NWS still provides a lot of useful information. Each regional NWS office now has its own home page and most of the field offices have or will soon have their own home pages (see the list on page 278). So chances are pretty good that you can locate a nearby NWS office's home page.

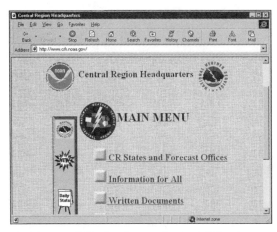

Each National Weather Service Regional Office has a home page as do many local Weather Service Offices.

Obviously, the more local the weather information is, the more useful it is.

Another good place to look is your state's land grant university's college of agriculture home page. Many land grant universities provide agricultural weather services on the Web. If your state's land grant university doesn't have weather, checkout other nearby state colleges and universities.

The University of Kentucky's Agricultural Weather Center, for example, is an excellent weather resource that provides agricultural weather for 12 states, not just Kentucky. The address for this site (which has many links to other weather sites) is **wwwagwx.ca.uky.edu**.

UK's Agricultural Weather Center was given a Four Star rating by Magellan, and a quick visit will show you why. The site has about everything you could want in the way of weather information for farming. Here, you'll find dew point maps, precipitation estimates and forecasts, daily soil temperatures and, yes, Doppler radar. You'll also see why weather information from local sources is the best kind and why having precise weather information can help you plan and manage your farming operation better and increase profits.

For example, the University of Kentucky tracks the number of degree days for various insects, like the alfalfa weevil, that impact the state's main field crops. Farmers don't even need to look for weevils in their alfalfa until something known as the "alfalfa weevil degree days" hits 190 or above (before that point, the insect can't mature). The University provides county-by-county maps that forecast when this pest and others hit the prescribed number of degree days and are likely to become a problem. This information lets farmers time pesticide applications with precision, increasing effectiveness, decreasing costs and helping protect the environment. For example, if the degree days in your county for the alfalfa weevil are 150 or less, then you don't have an alfalfa weevil problem—you probably can't find even one. UK's Ag Weather Center will report when the degree day number hits 180. At that point, you can start looking for the little pests, but should you apply pesticide right now? The answer is complex. The weather may not be right for pesticide application. You may also want to check the infestation level (how many bugs per square yard) and determine how big a problem you really have. You might also check the cost of the pesticide application compared to the level of infestation and value of the crop. It's possible to determine, from the infestation level, how much crop damage those little bugs will do, and since you already know what it's going to cost to spray for them, you can decide if it's worth it or if you should wait. For example, it doesn't make economic sense to spend $5,000 on a pesticide application to save $3,000 worth of corn. Yet, you can get all of this information from the Internet—and then make your decision (your extension agent can help with the calculations).

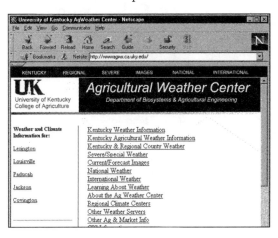

The University of Kentuck's Agricultural Weather Center sets the standard for Ag weather information. The address is **wwwagwx.ca.uky.edu**.

While we're partial to UK's Ag Weather Center (because we're located at UK), many other colleges of agriculture have ag weather, too (see page 278). If your state's land grant university doesn't provide a similar service, you should be able to patch together a reasonable substitute using information from various weather providers on the Internet—just remember to bookmark the addresses when you find them so you won't have to search for them again.

A third site that may help is a local television station's Web site. Many TV stations now have Web sites, and most have local radar, updated every 60 seconds or so (some are even live). Since Doppler radar is good out to about 160 miles, the television station doesn't need to be too close by to be useful to you.

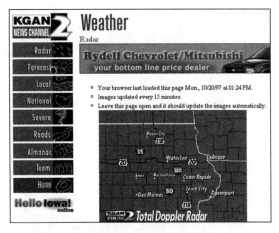

Live Doppler Weather Radar makes it easy to track storms. This shot is from the Web site of KGAN-TV in Cedar Rapids, Iowa. Many TV stations now provide "live" local radar coverage over the Web.

Most TV stations advertise their Web address at the end of their newscast. If you don't know the Web addresses for a TV station in your area, try **www.tvweather.com** for a list of all (or nearly all) TV stations that have weather sites on the Web. A good thing about using the Web site of a local TV station is that their Web server is probably not going to be overloaded when bad weather happens somewhere else in the world. This is especially important when you want to download a satellite map or Doppler radar, because images are larger and take much longer to download than plain text.

Admittedly, trying to find good sources of agricultural weather can be frustrating, but once you've found them (and bookmarked them), all you'll have to do to get the weather is point and click.

CHAPTER 9
AGRICULTURAL MARKET INFORMATION

Few things are more important to farming than fast access to market pricing information. Some farmers use sophisticated computer models and charts to determine the best time to sell (or whether or not to hedge), and some farmers just want to know the current local price for their particular crop. Better information makes for better decision making and improved profitability.

A few years ago, in a now famous study called *The Telegraph and the Structure of Markets in the United States*, Richard DuBoff showed how the expansion of telegraph lines in the late 1800's reduced the spread between the prices farmers were paid for their crops "at the farm" and the prices those crops actually sold for in major markets. Armed with the pricing information they could obtain by telegraph, farmers knew what their crops were selling for in Chicago, Minneapolis, and Topeka, and they were far more likely to get a fair price for those crops because of the information that came "over the wire."

Today, there are many ways to get market information, ranging from local newspapers and radio stations to satellite-based information systems like DTN FarmDayta, and, of course, you could always use the telephone and call the local stockyard or grain elevator. But if you have access to the Internet, you can quickly and easily obtain regional, national or even world commodities prices at virtually any level of detail, 24 hours a day, seven days per week. The Internet may have a similar effect today on farm profitability as the telegraph did a century ago.

FINDING LOCAL MARKET INFORMATION

Many local radio and TV stations, regional ag networks, and newspapers now have Web pages that cover agricultural pricing information (in addition to local news, weather and sports). The Web address is usually given at the end of the newscast or, in the case of a newspaper, is usually listed on the editorial page. Your "local" market may not be where you live, so you may have to hunt for the Web addresses for radio and TV stations or newspapers in the town or city where you actually market your crops and/or livestock. If you have trouble finding the Web address of a broadcaster or newspaper in a particular area, there are several sites that can help. A good place to search for local radio and television stations in your area is the Media Online Yellow Pages (**www.webcom.com/~nlnnet/yellowp.html**) which maintains a list of media resources on the Internet. For a list of newspapers with Web sites, try Newslink (**www.newslink.org**). It offers a variety of links to newspapers online. If you have problems locating a local site with the Media Online Yellow Pages or the Newslink site, you may have to resort to search engines (see Chapter 11) and search for the name of the newspaper or the call letters of the radio or television station.

OTHER SOURCES OF LOCAL MARKET INFORMATION

Many grain elevators and stockyards also have their own Web pages because they are finding that posting bid prices and updating their information online cuts down on the number of phone calls they receive every day. Customers also like the ability to go to a local grain elevator's or stockyard's Web site any time of day or night and get their market information. In addition, many of these elevator and stockyard Web sites maintain archives of past bid prices, so customers can get historical price information to use in charting market data and making sell or hedge decisions. You'll probably have to call to find out if they have a Web site.

One very good example of how local price information can (and should) be made available is the Brown County Coop in Hiawatha, Kansas. The Brown County Coop maintains a Web site (**www.bcca.net**) that lists their grain bid prices. Another grain elevator, Farmer's Co-op Elevator in Hemingford, Nebraska, also maintains a Web site (**www.bbc.net/co-op**) that lists their grain prices and has links to regional weather, too.

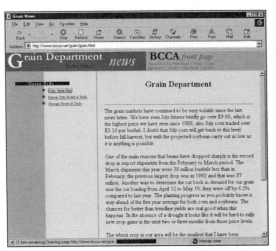

The Brown County Co-op in Hiawatha, Kansas, maintains grain news as well as links to bid prices.

Once you locate a good source of local ag market information, be sure to "bookmark" the site so you can go back to it without any trouble (see Chapter 7).

OTHER AG MARKET WEB SITES

TVA's Rural Studies Program at the University of Kentucky maintains a list of major market sites at **www.rural.org**. (A complete list of these sites can be found on page 216.) Obviously, one of our favorite places to go for regional information is the University of Kentucky. The University of Kentucky maintains USDA market price information on their gopher site (which simply means that you must type the **gopher://** before the rest of the address instead of **http://**). The University of Kentucky sites are:

USDA's Daily Cash Grain Reports from the University of Kentucky
gopher://shelley.ca.uky.edu:70/11/agmkts/market_wire/grain

USDA's Daily Future Grain Reports from the University of Kentucky
gopher://shelley.ca.uky.edu:70/00/.agwx/usr/markets/usda/MSGR711

USDA Market Wire News Reports from the University of Kentucky
gopher://shelley.ca.uky.edu/11/agmkts/market_wire

If you don't find what you are looking for on one of the University of Kentucky sites, try another land grant university or university extension site in your state. A list of land grant university Web sites can be found on page 179, and you can find extension sites on the Cooperative Extension System Information Servers Web site at **www.esusda.gov/statepartners**. If these methods don't work for you, try checking with your local extension agent or other farmers who are on the Internet.

Information from some universities may be found on a "Gopher" site instead of a Web site. You can get to any Gopher site with a regular Web browser like Netscape by typing **gopher://** (instead of http://).

PROFESSIONAL MARKET INFORMATION SERVICES

Many farmers and agribusinesses have been using the DTN or DTN FarmDayta satellite-based information services for years. These satellite-based information services are excellent resources, but in the past, they required an expensive satellite dish system that few smaller operators could afford. Today, DTN and DTN FarmDayta provide their information on the Internet for a small monthly fee. The Internet version of DTN FarmDayta (**www.dayta.com**) contains market and commodity prices, market and headline news, weather, futures options and charts, cash prices, chat groups and forums, as well as other services. The cost for the DTN FarmDayta Internet service is (at the time of this printing) $24.95 per month (on a month-to-month basis) or $240 for an annual subscription. For those of you north of the border, the Canadian version of DTN FarmDayta On-Line information is also available via the **www.dayta.com** site.

DTN FarmDayta information is now available over the Internet (for a fee). You don't need a satellite dish, just a computer and modem.

Although similar to the DTN FarmDayta site, DTN's main Web site (www.dtn.com) also contains new and used automobile information, financial services information, such as real time quotes and fixed income news, energy industry quotes and news, and news and price information for the produce industry. The point is that you can now get satellite-based ag information over the Internet, and since you don't need an expensive satellite system, the monthly fees are usually a lot lower.

NATIONAL AG MARKET INFORMATION

The following Web sites cover certain commodities and exclude others, so you'll have to check them out until you find the one that best fits your operation. Don't forget to bookmark the sites that you want to revisit. Also, check our TVA Rural Studies Web site at www.rural.org for the latest updated list of market sites.

AgriGator Ag Market News
http://www.ifas.ufl.edu/www/agator/htm/agmarket.htm

Agriculture Marketing Service (USDA)
http://www.ams.usda.gov/

Chicago Board Options Exchange
http://www.cboe.com

Chicago Board of Trade
http://www.cbot.com

Chicago Mercantile Exchange
http://www.cme.com

CNNfn - Commodities
http://cnnfn.com/markets/commodities.html

Farm Journal's Market Information
http://www.farmjournal.com

The University of Florida's *AgriGator* Web site is a great source of market information and other farming resources as well.

The Kansas City Board of Trade
http://www.kcbt.com/

The Livestock Marketing Information Center
http://lmic1.co.nrcs.usda.gov

NASS Reports from USDA
http://ag.arizona.edu/AREC/mnews/NASS.html

New York Cotton Exchange
http://www.nyce.com

New York Mercantile Exchange
http://www.nymex.com/

Successful Farming's Agriculture Online Markets
http://www.agriculture.com/markets/mktindex.html

CHAPTER 10
UNIVERSITY AND EXTENSION INFORMATION ON THE INTERNET

Researchers at universities around the world were the first people to use the Internet. (Actually, they were first because they *created* the Internet to help share data between researchers.) So it's only natural that universities, particularly *land grant* universities, are major sources of agricultural information on the Internet. Today, virtually all universities and colleges of agriculture have at least some information available on the Web, and many have comprehensive research, reporting, and extension information systems on the Net. This allows Internet farmers to have access to vast quantities of information that used to be time consuming to find or costly to obtain.

You can use these university Web sites to help in your business operations or everyday life. For example, what if you have questions about corn borer infestation after a late planted crop? You could go contact your local agricultural extension agent, but what do you do if it's 10:00 p.m. on a Friday evening and you would like to head for the co-op first thing in the morning to get the "right" pesticide? Well, you could get on the Web, surf to the nearest university's college of agriculture (or ag extension site) and see if they have any information online about the corn borer. You could also check with one of the agricultural chemical companies online—see page 129. Many times, the most up-to-date information about pests and pest control can be found on the Internet before it has had time to be published in hard copy. This is especially true of seasonal infestations or rapidly spreading outbreaks.

Here's how a typical search would work: Let's say you're a Kentucky farmer who grows corn. You begin to notice corn borer larvae starting to appear late in the growing season. You would first go to the main home page of the University of Kentucky's College of Agriculture (**www.ca.uky.edu**). From there, you would click on the **Entomology Department** link and look for information or articles explaining how to control the corn borer in Kentucky. If that pest is a particularly big problem this year, the department has probably issued news releases and bulletins. You could also check the College's Ag Extension links.

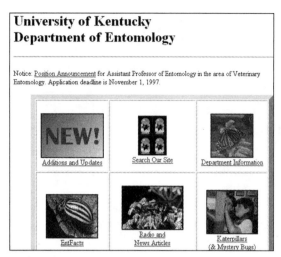

The University of Kentucky's Department of Entomology's home page can help provide solutions to those problems that are bugging you.

To continue with this example, by researching the information on UK's College of Agriculture site, you might learn that there is no reason to rush to the co-op tomorrow to get insecticide for the corn borer since insecticide control rarely works at the larval stage (because the larvae attack the base of the plant and not the leaves where most of the insecticide goes). Instead, the advice might be that your fields should just be harvested early to minimize crop damage. This research method sure beats playing phone-tag. (And, in this example, it saved additional time and money.)

FINDING UNIVERSITIES ON THE INTERNET

In order to find ag-related information from a university's Web site, you must first get to the university's main home page. (For a list of land grant universities on the Web, see page 179 or visit our Web site at www.rural.org.) Another way to find a list of universities is by using a search engine like Yahoo (see Chapter 11). Yahoo maintains a list of universities online at www.yahoo.com/Education/Higher_Education/Colleges_and_Universities. This site lists colleges and universities by country, so you can go to the United States link or anywhere else in the world to find out what kind of research information is available online. Local information is almost always the best, so check first with your state's land grant university or the ag college closest to you.

Once you find the university's main home page, you should see a link to the university's College of Agriculture Web site. Within the College of Agriculture's Web site, you may see several departments or categories of information. In some cases, you may need to search within the site (using "key words") to find what you want. For example, "whole farm" management information might be in the Department of Agricultural Economics or it could be in the Agriculture Business Department. The information about the corn borer might be found in the Entomology Department under "pest management," in the Agronomy

Department under "pest news" or "corn information," or even under Ag Extension Services Publications.

Once you have found a link to a site that covers your topic, click on it and look for publications and extension/research information online related to your topic. If you have searched three or four universities without success, then you may want to search for the e-mail address of an extension agent or university professor and contact him/her directly. In our experience, the ag faculty and staff at most universities are good about responding to e-mail questions (but keep in mind that the person you have contacted may be at a meeting, teaching classes, at field day, writing a paper, on vacation, etc.).

To give you yet another example, one of my neighbor's children found a small, newborn rabbit and wanted to "save it." The parents tried to keep it alive by feeding it cow's milk from the refrigerator. I don't know anything about rabbits, but I didn't think that cold cow's milk straight from the refrigerator was the right thing, so I got on the Web and searched for information about rabbit feeds. I used a search engine (Chapter 11) and found out about some research on "neonatal rabbit feeding" at Texas A&M University. I found the professor's name who did the research, got his home number, and called him on a Sunday night to find a way to try to keep the baby rabbit alive. It turned out that he was one of the nation's leading experts on feeding newborn rabbits and suggested that the parents go to a local pet store and buy kitten replacement milk to feed the baby rabbit. So, by using the Net to search for information, I was able to get in touch with one of the nation's experts and got the best advice Texas A&M had to offer about keeping a baby rabbit alive. Needless to say, the neighbors were impressed with my sudden expertise in "neonatal rabbit feeding."

The point of this story is that sometimes all you need is the <u>name</u> of someone to contact at a university about a specific problem or question. Instead of making long distance calls and running up your phone bill by being transferred all over the campus, go to the university's home page and see if they have a directory of employees listed by college or department. If so, you can look someone up in the directory and send them an e-mail message with your question. By the way, e-mail is preferred over phone calls, but the baby rabbit thing was an emergency.

An example of a typical faculty directory can be found at North Carolina State University's Web site: **www.ncsu.edu**. The home page has a link called "Directories." After you click on the **Directories** link you will get a page for "Colleges and Schools." On that list you will see a link to "Agriculture and Life Sciences." Next, click on that link and you will see a list of departments within the College of Agriculture. Most university home pages work this way.

For example, if you are looking for information on "whole farm" management, you may want to look in the Agricultural and Resource Economics Department. To find faculty information click on the **Faculty** link and you will find a list of all professors who work in that department.

To find out more information about the professors, just click on their name and you will get information such as area of specialty, office location, telephone number, fax number, and the all-important e-mail address. You should be able to find a research or extension professor or an assistant who specializes in the particular area you need information about. Again, as a courtesy, use e-mail to communicate with them, but only after you have carefully searched the university's site and are certain that the information you need is not there. Nothing is more frustrating that to constantly reply to inquiries when the information is there on the site and people are simply not making an effort to look for it. So, look first, and ask questions (via e-mail) later—when you are certain the information is not already posted on the site.

You can usually find contact information for faculty and staff members on the university's Ag college home page.

CHAPTER 11
SEARCH ENGINES

The Internet provides electronic access to thousands of computers, millions of people, and billions of pages of information. But to get to those computers, people and information, you have to know their location. More particularly, you have to know an *address*.

On the Internet, an address is known as a Uniform Resource Locator (URL). Once you know the URL of a document you want or an Internet site you want to visit, you type the URL into your Web browser's "Go to" window, and it will take you there. (Likewise, you must know a person's e-mail address before you can send e-mail to them.)

When you find the Uniform Resource Locator (URL) for a site you want, you must type it exactly. URLs are case sensitive. If part of the URL is in lowercase letters, use lowercase, and if parts are in uppercase letters, use uppercase. URLs frequently use characters like the forward slash (/), the tilde symbol (~) or the underline character (_). Be sure to use those same characters.

There are several ways to find the URLs for the information you want. First, there are printed lists of URLs (like the one in this book). Printed lists are convenient and easy to thumb through. But this convenience comes at a price: printed lists (including the one in this book) eventually go out of date. To help overcome that limitation, we maintain an up-to-date online version of the lists in this book. To get an update, simply visit our Web site **www.rural.org**. You can either click through menus to find what you want or you can type in key words and search for sites that match.

Another way to locate information is to go to a home page that relates to the subject matter that interests you. Home pages (and other sites on the Web) usually contain hypertext links that allow you to move from one Web site to another just by pointing and clicking on the highlighted text (usually a different color from the main text or underlined). And unlike printed lists, hypertext links can be easily changed, so they are usually more up-to-date.

In addition to having links to other, similar sites, some sites also have built-in search engines that can help you find related materials. For example *Farm*

Journal Today (www.farmjournal.com) has a built-in search engine called AgFinder® which can quickly help you locate agricultural resources on the Web. You may have noticed that URLs are now being included in news articles and in magazine and newspaper ads and television shows. Interesting sites are also frequently posted to newsgroups and online forums.

Next in line are the search engines. These are computer programs on the Internet that hold vast amounts of information about what's on the Web. Some search engines are better than others, depending on what you are looking for. Some search engines are devoted to a single topic like farming. One is AgriSurf (www.agrisurf.com). Another good one is run by Canada's University of Guelph (pronounced "gwelf"). The address is www.caffine.ca. Although it's primarily aimed at Canadian farming, it contains a lot of information that would be useful to farmers anywhere.

Once you find a site you plan to visit again, you should store its URL so you won't have to type it again and again. You can do this by adding the site to a list of important or frequently visited sites. In Netscape, this custom list of URLs is called Bookmarks, while Internet Explorer calls them Favorites. When you are at a Web site and you want to store its URL, open the Bookmarks window and click Add Bookmark. (Or in Explorer, go to the toolbar at the top and click on Favorites and then Add Page to Favorites). The next time you want to go back to that site, open the Bookmark (or Favorites) window, and pick the site you want from the list. You'll go straight there, and you won't have to type (or remember) all those URLs.

Further down the list are the general purpose search engines like Yahoo, Lycos, and InfoSeek. Search engines let you take an active role in looking for what you want. You don't need any special software, just your Web browser. The search engine itself is on a computer located somewhere out on the Internet.

One problem with general purpose search engines is that it is possible, and even likely, that a really long document will contain at least one of the words you're searching for without really relating to the subject you are interested in. For example, if you type in the word "Beef" you might get several thousand "hits" but most will be about cooking beef, not about raising cattle. Some search engines will let you skip documents unless the search terms are close together or in a certain order. This technique helps you distinguish, for example, between documents about ant problems on the farm versus ant farms—an important distinction.

USING SEARCH ENGINES

Many Web browsers are already set up with links to the most popular search engines. In Netscape, for example, you'll find a list of search engines under the Net Search button.

Each search engine uses a different method for conducting a search. Some search engines let you use "and" or "or" to tie terms together, but others might use the plus sign (or some other method). Most search engines will let you enter the words you want to search for in quotes, which usually forces the search engine to find only those words in that exact order. For example if you entered "chicken manure" (in quotes) the search engine would list only those sites that had a reference to *chicken manure*, rather than list all the sites that had the word *chicken* (including recipes for *chicken italiano*) and all the sites that had the word *manure* (which may include old speeches by President Truman, whose wife reportedly tried to make him use the word *manure* instead of other, more colorful references).

Search engines tend to fall into categories. Some are fast and easy to use, but don't contain much depth. Others are harder to use but are more thorough. As a rule of thumb, try a fast, easy-to-use search engine for your first attempt. If you can't find what you want, you can move on to more comprehensive search engines.

Search engines can be quirky and may not return the results you expect. Sometimes you know something is out there, but the search engine can't find it. They are not very good at guessing, and if the creator of the document you're searching for didn't include the exact keywords you are looking for, the search engine will have trouble finding the site.

Most search engines are easy to use, but some use highly structured syntax rules to direct the search. These rules are great in theory, but they don't always work as advertised. In practice, it may take several attempts at getting the syntax right for the search engine to give you exactly what you want. These rules also vary from search engine to search engine. Using the InfoSeek syntax won't work when you're in Yahoo, for example. Most search engines have a Help or Tips section which you can print out (offline, of course) and keep as a reference.

If you have trouble thinking up appropriate search terms or keywords, you should make your best guess and look at the results you get for additional clues about how to proceed. Look at some of the pages you found for other key words to try. You might be a hog producer, but some sites might list their information under pigs or pork, so keep looking.

YAHOO

http://www.yahoo.com

Yahoo is one of the most popular search engines on the Net. It's easy to use and usually pretty fast. Don't expect an in-depth search with Yahoo, but if you're looking for topical information of general interest, Yahoo probably can find out something about it.

When you first start Yahoo, you'll see lists of subjects including Business, Government, Computers, and Science. These are prepackaged lists of frequently requested information. It's often faster and easier to use these lists than to search for specific terms, but Yahoo will also let you type the name of a subject and search for anything you want.

Yahoo is a very good resource for finding local Internet Service Providers.

INFOSEEK

http://www.infoseek.com

InfoSeek is more powerful than Yahoo, but this power comes at a price—it requires a little more effort on your part in specifying how InfoSeek should look for information. See InfoSeek's Helpful Tips section for more information about using proper syntax and other methods to refine your search.

LYCOS

http://www.lycos.com

Lycos is a comprehensive search engine developed at Carnegie Mellon University. One of the best features of Lycos is the Search Options form. You tell Lycos how to match the terms you're searching for, how close a match you want, and how many results you want per page. For example, if you told Lycos to match all terms in a search for beef respiratory diseases, it would return only those pages that had all three terms. Or you could tell Lycos to match any term; in this case a search for beef cattle would show you everything Lycos found with the words beef, cattle, or even beef cattle. This is a handy feature if you aren't sure whether you should be looking for swine, pigs, hogs, or pork, for example.

▼ *Chapter 11*

MORE SEARCH ENGINES

Accufind Net Locator
http://nln.com

AgriSurf (devoted to farm and agriculture topics)
http://www.agrisurf.com

The All-in-One Search Page
http://www.albany.net/allinone

Alta Vista Query
http://altavista.digital.com/

Bigfoot Search (e-mail addresses)
http://www.bigfoot.com

Deja News (for Internet Newsgroups)
http://www.dejanews.com

Excite
http://www.excite.com

Four11 Directory Services (an Internet White Pages)
http://www.four11.com

HotBot Search
http://www.search.hotbot.com/index.html

Magellan Internet Guide
http://www.mckinley.com

Web Crawler
http://www.webcrawler.com

The Who/Where Search Engine (great for e-mail and phone numbers)
http://www.whowhere.com

Printed listings get out of date quickly. We continually search for farm-related sites and will post the results from time to time on our home page at www.rural.org. New links that we add will have a "NEW" icon beside them.

Also, Farm Journal keeps a close eye on all subjects related to farming, including the Internet. Look for updates there, too. The address is: www.farmjournal.com (they also have links to other Web sites that can help you in your search).

CHAPTER 12
FARM JOURNAL'S WEB SITE

With thousands and thousands of farming-related Web sites, you'd think it would be hard to know where to start. Well, it isn't. The best starting place on the Internet for farming information is *Farm Journal Today*, produced by *Farm Journal*—one of the most trusted names in agriculture for more than 120 years. In fact, you may be able to find everything you need right there in one neat package.

In a recent Cybertimes column, *The New York Times* reported "the venerable magazine of rural life, *Farm Journal*, has a sleek, rich Web site with articles from the print edition, business planning information and links." *The New York Times* also praised *Farm Journal*'s work in creating an interactive dialogue with farmers, commenting that "it's a great site that lives up to the Net's two-way promise." Need we say more?

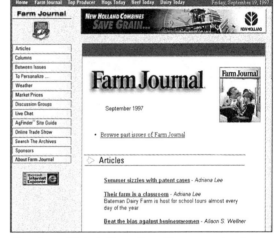

Farm Journal Today's Home Page. Due to continual improvements, this welcome screen may look a little different when you log on.

The Web address for *Farm Journal Today* is **www.farmjournal.com**. You can get there by simply typing the address in the "Go to" (or address) box on your browser, or, better yet, you can make *Farm Journal Today* your full-time home page. When you start your browser, it probably goes to Netscape or Microsoft's home page, which is good if you need to update your software, but for most of us, it can just be a waste of time. Instead, tell your browser to start at some place really useful, like *Farm Journal Today*. (To learn how to make *Farm Journal Today* your home page, see Chapter 7.) If you set your browser's home page to *Farm Journal Today*, you'll be "in the know" as soon as you connect to the Net.

If this is your first visit to *Farm Journal Today*, you'll notice a series of boxes in the frame on the left-hand side of the main screen. You can click on any one of the boxes to quickly go to information you want.

The first box lets you personalize *Farm Journal Today* based on where you farm and what type of operation you have. For example, if you take a few moments the first time you visit to personalize *Farm Journal Today*, the next time you load *Farm Journal Today*, and select, say, Weather, you'll get the weather for your local area without having to click through maps of the whole country. The same with Market Prices. Once you've personalized *Farm Journal Today* for your farming operation, you can click on Market Prices, and go straight to the market information for the commodities that interest you.

Of course, you can click on Weather and Market Prices without first personalizing the home page, but then you'll always have to tell it where you are and what crop or livestock information is important to you. Did we mention that this service is free?

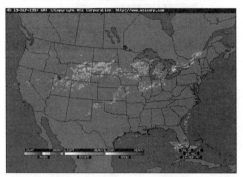

Farm Journal Today lets you personalize the site to suit your type of farming operation and location.

The **Weather** box has links to all the major sources for agricultural weather, including NEXRAD weather radar and a list of TV stations that put their local Doppler radar on the Web (very handy).

Click on **Market Prices** and you'll get the latest quotes from the Chicago Board of Trade, the Chicago Mercantile Exchange and other major trading centers, in addition to an assortment of cash market prices.

The next two boxes allow you to participate in discussions about agricultural issues. Click on **Discussion Groups** and you'll get a list of current topics (called threads). You can click on a topic and read what other farmers are saying, and you can add your own two cents worth, too. The topics under discussion change regularly and new ones are being added continuously.

The **Live Chat** button lets you engage in live online discussions with your colleagues in one of several chat rooms. Chat rooms are different from Discussion Groups in that they are "live"—all of the participants are online at the same time, typing comments back and forth.

When you first click on **Live Chat**, you'll get a message about which chat areas are active. If one is empty, then no one is around to chat with. But, day or night, you'll find lively discussions on a variety of farming topics and form a new community of friends in the process.

MONDAY NIGHT CAMPFIRE

Most Monday nights, beginning at 7 p.m. Central Time, *Farm Journal* editors invite you to sit around the electronic campfire and "chew the fat." Like the chat rooms, this is a live online discussion. But unlike other chat rooms, the Monday Night Campfire is a scheduled event and often includes guests as well as participation by *Farm Journal* editors.

A number of other scheduled events are promoted ahead of time at the site. One is "First Tuesday" which is held on the first Tuesday of each month. "First Tuesday" is a live chat series that puts producers in direct contact with top-ranking executives of major agri-business firms to share questions, ideas, and information. This is a rare opportunity you wouldn't otherwise have, and a direct benefit from being on the Internet.

Farm Journal's Monday Night Campfire is an opportunity to sit around the warm glow of the computer screen and "chew the fat" with other farmers.

OTHER SITE FEATURES

The other boxes along the left-hand side let you search for ag-related Web sites (using tools such as AgFinder™) or you can browse current or back issues of *Farm Journal* publications or get additional product information from advertisers (sponsors).

AgFinder™ is an extremely important time-saving navigation tool. AgFinder™ lets you search for and link to other business-related sites you may not know about (as do the banner ads promoting the products and services of commercial companies with their own Web sites).

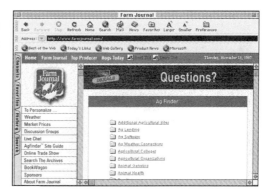

Can't find what you are looking for? Let *Farm Journal Today*'s AgFinder™ find it for you!

The main screen has two other very useful features. The first is "Between Issues." Here you'll find reports on new business developments from the folks at *Farm Journal* since the latest printed issues were mailed. The "Feature Articles" section highlights some key stories from *Farm Journal's* entire family of publications including *Farm Journal, Top Producer, Beef Today, Dairy Today* and *Hogs Today*. If you want to see more from any of these publications, simply click on the bar at the top of the page for the name of the publication you want, then browse through Articles/Columns/Between Issues choices.

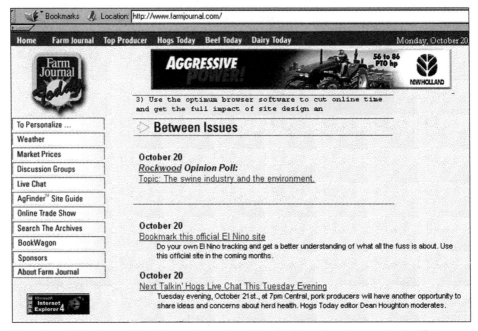

Farm Journal Today (**www.farmjournal.com**) provides the latest farming news in the "Between Issues" section. You can also look at the lead stories from the entire family of *Farm Journal* publications including *Farm Journal, Top Producer, Beef Today, Dairy Today* and *Hogs Today*. *Farm Journal Today* is your "one stop shopping" source for the latest farming news and information.

LEARNING TO CHAT ONLINE

Participating in live online discussions can be a little strange until you get used to how they work. It's important to realize that it takes time for the other people to read what you wrote and to type a response. It may take several minutes. Be patient. A short delay doesn't mean that no one is interested in what you just said. Some people just type slower than others!

Second, bend those rules of etiquette your parents taught you, like "keep it short" and "don't interrupt." These kinds of rules were meant for face-to-face conversations, not the Internet.

Since it takes awhile to read and respond to messages, don't try to be too "conversational"—the lag time is much too long to try to conduct snappy back and forth dialogue. Instead, when you send a message, explain yourself fully and include all pertinent information. You've got to put enough information in your message without being "long winded" so readers can understand your comments.

Another thing that will take some getting used to is having several conversations or threads going on at the same time. You might see a message on one topic, and while you are typing a response, several messages about different subjects will appear "out of order." This is normal. There will be several, usually unrelated, threads scrolling up the screen at the same time. Disregard the ones that don't interest you. It may seem jumbled at first, but you'll get used to it. You don't have any control over when your message gets posted, so when it does show up, you're not really "butting in" to someone's conversation.

Lastly, don't fret over typos and spelling. When they are online, even *Farm Journal* editors miss the keys from time to time.

PART II
INTERNET ADDRESSES

	Page
AGRICULTURAL RESOURCES	123
HOT Agricultural and Farming Links	123
Agricultural Companies Online	129
Agricultural Magazines Online	142
Agricultural Organizations Online	143
Agricultural Related Newsgroups Online	146
Agricultural Software Online	147
Alternative Agriculture Sites	148
Crop Resources	149
General Crop Links	149
Apples	154
Berries	155
Canola	156
Citrus Fruits	157
Coffee	158
Corn	160
Cotton	161
Hay and Pasture	162
Industrial Crops	164
Other Fruits	164
Peanuts	165
Rice	165
Small Grains	168
Sorghum	169
Soybeans	170
Tobacco	171
Tubers	171
Vegetables	171
Entomology	173

Farmers Online	174
Forestry	178
Land Grant Universities	179
Livestock Resources	200
General Livestock Links	200
Beef Cattle	203
Dairy	206
Goats	207
Horses	208
Other Livestock	210
Poultry	211
Sheep	212
Swine	213
Mailing Lists On the Internet	215
Management and Marketing	215
Market and Price Information Sites	216
Pesticides	219
Precision Farming	220
Soil and Water	223
State Departments of Agriculture	225
Turf Management	228
Wildlife	230
ARTS AND SCIENCE	230
BUSINESS AND FINANCE	232
COMPUTER RESOURCES	238
Computer Magazines	238
Computer Retailers	240
Computer Software	241
E-Zines - Electronic Magazines	244
Hardware Companies	246

IBM Compatible Computer Resources	248
Internet Service Providers	248
Macintosh Resources	249
ECONOMIC DEVELOPMENT	250
EDUCATION AND REFERENCE	254
GOVERNMENT	255
HEALTH	258
HISTORY	260
HOME AND GARDEN	261
LAW	263
NATURAL RESOURCES	264
NEWS	265
SEARCH ENGINES AND LISTS	268
SPORTS AND RECREATION	271
STATE INFORMATION AND UNIVERSITY RESOURCES	272
TAX INFORMATION	273
TELECOMMUNICATIONS	274
TELEPHONE COMPANIES	276
TRAVEL	276
WEATHER	278
USA Weather	278
Regional Weather	282
National Weather Service Field Offices	284
Canadian Weather	291
International Weather	291

AGRICULTURAL RESOURCES

HOT AGRICULTURAL AND FARMING LINKS

Ag Agent Handbook
http://hammock.ifas.ufl.edu/text/aa/31541.html
Information about everything from apples to swine to control of small animals resides at this site.

Ag Links from the Gennis Agency
http://www.gennis.com/aglinks.html

AgJobs USA
http://www.agjobsusa.com

Agricultural Genome World Wide Web Server
http://probe.nalusda.gov:8000/
An integrated system for agricultural genome analysis.

Agriculture Engineering Information from the University of Florida
http://hammock.ifas.ufl.edu/text/aa/39592.html
Contains information about agricultural safety, environmental requirements for greenhouses, and farm machinery.

Agriculture Network Information Center
http://www.agnic.org
AgNIC is a distributed network that provides access to agriculture-related information, subject area experts, and other resources.

AgriGator - Index of Agricultural and Related Information
http://gnv.ifas.ufl.edu/WWW/AGATOR/HTM/AG.HTM
Everything you ever wanted to know about agriculture can be found through this site. It is constantly updated.

Agrinet
http://agrinet.tamu.edu/
AgriNet, a service of the Texas A&M Agricultural Program, was developed to provide a single starting point for all agricultural resources on the Internet.

AgriSurf
http://www.agrisurf.com
Agricultural search engine.

Agriweb
http://www.ruralnet.com.au/AgriWeb/
This is an excellent rural life and agriculture Web site. It has predominantly Australian information, but includes U.S. market reports.

Agview Cool Sites
http://www.agview.com/ag/ag_cool.cfm

Biological, Agricultural, and Medical INFOMINE
http://lib-www.ucr.edu/bioag/
Links and information including library resources.

Center for Animal Health and Productivity (CAHP)
http://cahpwww.nbc.upenn.edu/

Chicago Board of Trade
http://www.cbot.com
The CBOT offers information on producers, prices, and market information in AgriMarket and MarketPlex.

Chicago Mercantile Exchange
http://www.cme.com/
Futures, options, and price information (primarily livestock).

Commodity Traders Advice
http://infomatch.com/~adas/adv.html
This site contains ag weather and statistics, pre-opening comments, end of day quotes, and downloadable data and charts.

Cyberfarm
http://w3.ag.uiuc.edu:80/infoag/cyberfarm/

Don't Panic, Eat Organic
http://www.rain.org/~sals/my.html

DTN FarmDayta Online
http://www.dayta.com

EcoNet
http://www.econet.apc.org/econet/

Environmental Organization Web Directory
http://webdirectory.com/Science/Agriculture

Extoxnet
http://www.oes.orst.edu:70/1/ext
This site from the Oregon Extension Service includes pesticide information.

Farm and Resource Economics
http://hammock.ifas.ufl.edu/text/aa/39593.html

Farm Credit Services Mainstreet USA
http://www.mainstreet.com

Farm Journal Today
http://www.farmjournal.com/

Florida Agricultural Information Retrieval System (FAIRS)
http://hammock.ifas.ufl.edu/

Food and Agricultural Policy Research Institute
http://www.fapri.missouri.edu/
In studies ranging from the farm to the international marketplace, FAPRI uses comprehensive data and computer modeling systems to analyze the complex economic interrelationships of the food and agriculture industry.

Food and Agriculture Organization
http://www.fao.org/
This site has the World Agricultural Information Centre (WAICENT) publications update, information on workshops and seminars, plus a search facility.

Georgia Tech's Agricultural Technology Research Program
http://www.gtri.gatech.edu/ag-research/
This page contains links to information about safety, outreach, automation, and environmental programs. This site also maintains online publications.

Growmark, Inc. List of Agriculture Links
http://www.growmark.com/sites/web.html

IICA In Canada
http://www.iicacan.org/
Updated information on IICA's activities in Canada and offers a specialized site with links to what is happening in the world of agriculture and agribusiness.

Illinois Extension's VISTA Electronic Information Library
http://spectre.ag.uiuc.edu/~vista
VISTA is an electronic publishing project of the Illinois Cooperative Extension Service and the College of Agricultural, Consumer and Environmental Sciences.

Information Services for Agriculture
http://www.aginfo.com/
The ISA has weather and market information geared toward the agricultural community.

The Internet Services List for Agriculture
http://www.spectracom.com/islist/

KRVN-AM Farm Radio
http://www.krvn.com/
KRVN serves rural Nebraska, Kansas, Northeast Colorado and Southeast Wyoming farmers and ranchers with agricultural and educational information.

National Agricultural Library
http://www.nalusda.gov/
NAL is an international source for agriculture and related information, and provides access to NAL's many resources and a gateway to its associated institutions.

National Center for Agricultural Law Research and Information
http://law.uark.edu/arklaw/aglaw/
The NCALRI conducts research and analysis and provides up-to-date information to farmers and agri-businesses, attorneys, community groups, and others confronting agricultural law issues.

National Weather Service State Forecasts
http://iwin.nws.noaa.gov/iwin/textversion/states.html (text version)
http://iwin.nws.noaa.gov/iwin/iwdspg1.html (graphics version)
The Interactive Weather Information Network provides detailed state by state forecasts, as well as information on current weather watches and warnings.

NetVet
http://netvet.wustl.edu/vet.htm
Includes all kinds of information related to veterinary science, including searches, links related to household animals and livestock, lists of veterinary schools, and much more. Includes a What's New calendar of on-line animal related sites.

▼ *Part II*

Not Just Cows
http://www.snymor.edu/~drewwe/njc/
Wilfred Drew has compiled this list of Internet/Bitnet resources for agricultural information.

Ohio State University's Cyber-Farm
http://www.ag.ohio-state.edu/~farm
The Cyber-Farm is a link-filled, graphically oriented Web site that features sections on woodlot management, crops, wind and other alternative energy, farm machinery, grazing, pond management, home safety, and much more.

Pest Control Information
http://hammock.ifas.ufl.edu/text/aa/39594.html
Pest control data from the Florida Agricultural Information Retrieval System. Includes information on pesticide registration, news and training programs.

Rural Internet Access,, Inc. Home Page
http://www.ruralnet.net

Southwest Rural and Agricultural Safety
http://ag.arizona.edu/agsafety
Safety information, links and suggestions.

Today's Market Prices
http://www.todaymarket.com
Provides wholesale fruit and vegetable market prices from the United States, Canada, Mexico, Europe, Asia and Latin America, classified by product, terminal, varieties and sizes.

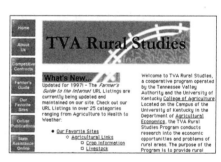

TVA Rural Studies
http://www.rural.org
This site focuses on rural economic growth ideas and opportunities. Includes links to valuable resources for everyone in the rural community, from farmer to entrepreneur, from teacher to mayor.

United States Department of Agriculture - USDA
http://www.usda.gov
This site provides detailed information about the U.S. Department of Agriculture, as well as links to USDA's individual agencies.

University of Kentucky Agricultural Weather Center
http://wwwagwx.ca.uky.edu
Maintained by the UK College of Agriculture, this site provides detailed agricultural weather and climate information for Kentucky, the United States, and the world.

US Farm Report
http://www.usfronline.com

USDA Animal and Plant Health Inspection Service - Index of Searchable Terms
http://www.aphis.usda.gov/oa/srchtrms.html
This site provides you with an index of searchable terms about APHIS and APHIS programs.

USDA Economic Research Service
http://www.econ.ag.gov
The Economic Research Service (ERS) is the economic research agency of the U.S. Department of Agriculture. ERS functions include research, situation and outlook analysis, staff analysis, and development of economic and statistical indicators in five areas.

USDA National Agricultural Library
http://www.nal.usda.gov
The National Agricultural Library is a major international source for agriculture and related information. This Web site provides access to NAL's many resources and a gateway to its associated institutions.

Virtual Virginia Agricultural Community Homepage
http://www.ext.vt.edu/vvac/
The Virtual Virginia Agricultural Community brings together Virginia's diverse agricultural industry into an electronic "virtual community" describing and supporting the state's agriculture.

The Weather Channel
http://www.weather.com
The Weather Channel online site provides local weather forecasts for hundreds of major cities across the United States, as well as business travel, ski, and tropical forecasts.

Weather Information at the University of Illinois at Urbana-Champaign
http://www.atmos.uiuc.edu/weather/weather.html
This is a list of popular weather and climate resources, available from the University of Illinois, Department of Atmospheric Sciences.

World Wide Web Virtual Library of Agriculture
http://ipmwww.ncsu.edu/cernag/cern.html
This site has agriculture links to any agricultural topic that interests you.

Yahoo's Agriculture Listing
http://www.yahoo.com/Science/Agriculture/
Includes an exhaustive list of agricultural information and links to other sites.

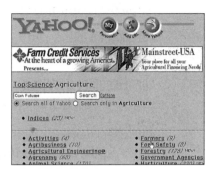

AGRICULTURAL COMPANIES ONLINE

ABS Global, Inc.
http://www.absglobal.com/

Aerotech, Inc. Agricultural Ventilation Systems
http://www.aerotech-inc.com

Ag Services of America
http://www.agservices.com

AGCO Farm Equipment
http://www.agcocorp.com

Ag-Decisions, Inc.
http://www.agdec.com

AgrEvo Liberty Link
http://www.agrevo-usa.com

Agri Motive Products, Inc.
http://www.ag-electronics.com

Ag-West Biotech Inc.
http://www.lights.com/agwest/

AgriBiz
http://www.agribiz.com/

Agricomm Agricultural and Equine Web Communication
http://agricomm.com/agricomm/

Agriculture.net
http://www.agriculture.net/

Agri-Financial Services
http://www.clickads.com/agrifinancial/

AgriHelp
http://spiderweb.com/ag/

Agri-Marketing
http://www.agrimark.com/

Ag-Mart Online
http://204.91.84.90
Web page development company specializing in Ag-related Web sites.

AgriMotive Products, Inc.
http://www.citznet.com/~vanguard

AgriOne: The Internet Agricultural Marketing Company
http://www.agrione.com

Alabama Farmers Cooperative
http://www.afc.com/

Albert Lea Seed House
http://www.alseed.com

Alberta Pool Online
http://fis.awp.com/

Allendale Inc.
http://www.allendale-inc.com

Allflex, Inc.
http://www.allflexusa.com/

American Agrisurance, Inc.
http://www.amag.com

American Ag-Tec International, Ltd.
http://www.ag-tec.com/

American Cyanamid
http://www.cyanamid.com

American Gelbvieh
http://www.gelbvieh.com

American Turbine Pump Co.
http://interoz.com/atp/atphome.htm

Amoco Lubricants
http://www.amoco.com/lubes

Ampac Seed Company
http://www.ampacseed.com/

Andersen-Balk Equipment Company
http://www.abguidance.com/

ARBICO - Arizona Biological Control Inc.
http://www.usit.net/hp/bionet/ARBICO.html

hot site

Archer Daniels Midland
http://www.admworld.com

Arctic Cat
http://www.arctic-cat.com

Asgrow Seed Technology
http://www.asgrow.com/

Ashtech
http://www.ashtech.com

Automatic Water Watcher
http://www.fishnet.net/~jackh/

BASF Ag Products
http://www.basf.com/index.html

Bayer Animal Health
http://www.uscrop.bayer.com

Bayer Corporation Homepage
http://www.bayer.com/bayer/english/0000home.htm

Bethurum Research and Development
http://www.bethurum.com/

Boehringer Ingelheim Animal Health, Inc.
http://www.boehringer-ingelheim.com/biahi/

BraMar Cattle Company
http://www.teleport.com/~bradp/

Brillion Farm Equipment
http://www.brillionfarmeq.com/

Broyhill Company
http://www.siouxlan.com/broyhill/

Campbell Seed, Inc.
http://www.campbellseed.com

Cargill Hybrid Seeds
http://www.cargill.com/seed

Carhartt
http://www.carhartt.com

Case Corporation
http://www.casecorp.com

Caterpillar
http://www.cat.com

Cattleco Internet Livestock Market
http://www.cattleco.com/

CattlePro Bowman Farm Systems Inc.
http://www.well.com/user/mason123/beef/cattlepro.html

Chevrolet Motors
http://www.gmc.om

Ciba Specialty Chemicals
http://www.cibasc.com

Claas of America
http://www.claas.com

Clark Consulting International Inc.
http://www.agpr.com/consulting/index.html

CNA Agriculture
http://www.cna-agtech.com

Colonial Home Page
http://www.colonialinc.com

Communication Systems International
http://www.csi-dgps.com
Specializes in differential GPS technologies.

Cozinco Sales, Inc.
http://www.cozinco.com

Daugherty, Inc.
http://www.daughertyinc.com

Deere and Company
http://www.deere.com

DEKALB Genetics Corporation
http://www.dekalb.com

Dickinson Cattle Company Homepage
http://www.texaslonghorn.com/

Dish Network
http://www.dishnetwork.com

Dodge Truck/Chrysler
http://www.4adodge.com/ram/index.html

Dow Chemical Company
http://www.dow.com

DowElanco
http://www.dowelanco.com

DTN FarmDayta Online
http://www.dayta.com

Duffield Real Estate Company
http://www.ruralappraiser.com/

Dupont
http://www.dupont.com/

Eastern Breeders Inc.
http://casper.ilms.com/ebi/

Ecostat, Inc.
http://lakeland.tsolv.com/~ecostat/

Elanco/Rumensin
http://www.elanco.com/Products.html

ESRI (GIS)
http://www.esri.com

Farm Credit Services - Mainstreet USA
http://www.mainstreet-usa.com

Farm Direct Marketplace
http://www.farmdirect.com/
This is an Internet farmer's market made up of real farmers.

Farm Works
http://www.farmworks.com/

Farmer's Insurance Group
http://www.farmersinsurance.com/

Farmer's Software Association
http://www.farmsoft.com/

Farmland Industries, Inc.
http://www.farmland.com

Farms Unlimited (Mort Wimple Associates Real Estate)
http://www.albany.net/~wimple

Fast Finder Online Equipment Directory
http://www.fastfinder.com

Fisher Scientific Internet Catalog
http://www.fisher1.com/index.html

Fitzpatrick Cattle Company
http://rfitz.com/cattle/

FMC Corporation
http://www.ag.fmc.com

Ford Motor Company
http://www.ford.com

Garst Seed Company
http://www.garstseed.com

General Motors
http://www.gm.com

Global Agricultural Biotechnology Association
http://www.lights.com/gaba

Golden Harvest Seeds
http://www.goldenharvestseeds.com

Goodyear
http://www.goodyear.com

Grand Laboratories
http://www.grandlab.com

Great Lakes Hybrids
http://www.glh-seeds.com

Greenleaf Technologies
http://www.wild.net/greenleaf/

Growmark
http://www.growmark.com

The GSI Group
http://www.grainsystems.com

Gustafson, Inc.
http://www.gustafson.com

Hagie Sprayers
http://www.hagie.com

Hawkeye Steel Products, Inc.
http://www.hawkeyesteel.com

Hay Exchange
http://www.hayexchange.com

Helena Chemical
http://www.helenachemicals.com

Hi Tech Detergents, Ltd.
http://www.hitechdeterg.co.nz/

Implement and Tractor Online
http://www.ag-implement.com

InfoHarvest Homepage
http://www.infoharvest.ab.ca

Information Services for Agriculture - ISA
http://www.aginfo.com/

InterUrban WaterFarms
http://www.viasub.net/IUWF/2.html

Inverizon International Inc.
http://www.inverizon.com

J.R. Simplot Company
http://www.simplot.com

Kawasaki Motors Corp.
http://www.kawasaki.com

Kello-Bilt Inc.
http://www.kello.reddeer.net

Keystone Steel and Wire
http://www.redbrand.com

Kverneland Agricultural Implements
http://www.kverneland.com

Land-O-Lakes
http://www.cnxlol.com

List of Agricultural Corporations Online
http://www.agric.gov.ab.ca/aginf06.html

Lonza, Inc.
http://www.lonza.com

Lowe's Home Improvement Centers
http://www.lowes.com

Magic Green Corporation
http://www.magicgreen.com

Maine-Anjou
http://www.maine-anjou.org

The Market Place / Farmer's Market Entrance
http://www.tastefulideas.com/marketplace/mpfront.htm

Monsanto
http://www.monsanto.com

MoorMan's Feeds
http://www.moormans.com

M-Tech International, Inc
http://ourworld.compuserve.com/homepages/Mtech_International/

Mycogen Seeds
http://www.mycogen.com

Nederlandse Verwarmings Maatschappij
http://www.caiw.nl/~dop/nvm/eng.htm
Greenhouse designers from the Netherlands.

New Holland Tractors and Equipment
http://www.newholland.com/na
Chosen as Best Website of 1997 by the National Agri-Marketing Association.

Northrup King - NK
http://www.nk.com

Northwest Precision Ag.
http://www.srv.net/~nwpag/NPA.html

Novartis (Merger of Ciba & Sandoz)
http://www.novartis.com

Omnistar DGPS Service
http://www.omnistar.com

Orange Enterprises, Inc.
http://www.orangesoftware.com

PellComm Online Marketplace
http://pellcom.com

Pennsylvania Farm Bureau
http://www.fb.com/pafb/

Pfizer
http://www.pfizer.com

Piedmont Farm Credit
www.agfirst.com/piedmont

Pilatus, Inc.
http://www.hydroculture.com

Pioneer Hi-Bred International
http://www.pioneer.com/

Pittsburg Tractor (Massey Ferguson Dealer)
http://members.aol.com/pittsburg/

Precisionag.com
http://www.precisionag.com

Pro Farmer Online
http://www.profarmer.com

Proven Seed Online
http://www.provenseed.com/

Purina
http://www.purina.com

Rhino Linings
http://www.rhinolinings.com

Rhone-Poulenc
http://www.rhone-poulenc.com

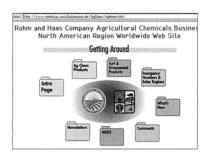

Rhone-Poulenc Ag Company
http://www.rp-ag.com

Rockwell Agricultural Systems
http://www.cacd.rockwell.com/bus_area/ag_sys/

Rohm and Haas Overview of the Ag Chemical Business
http://www.rohmhaas.com/businesses.dir/AgChem/AgHome.html

Schering-Plough
http://www.nuflor.com

Silver Creek Feeders, Inc.
http://www.silvercreekfeeders.com

Simmental
http://www.simmgene.com/asawebpg.htm

Soil Technologies Corp.
http://www.fairfield.com/soiltech

Southern States Cooperative, Inc.
http://www.sscoop.com

SoyPLUS - The Gold Standard
http://www.soyplus.com

Spectrum Technologies, Inc.
http://www.specmeters.com

SST Development Group, INC.
http://www.sstdevgroup.com
GIS and remote sensing technologies.

Stevenson-Basin Angus
http://www.stevensonbasin.com

Stine Seed Company
http://www.stine.com

Sunflower Manufacturing
http://www.sunflower-mfg.com/sun.html

Sunseeds
http://www.sunseeds.com

Suzuki
http://www.suzukicycles.com/quking.htm

Terra Industries
http://www.terraindustries.com

Titan Wheel International, Inc.
http://www.titan-intl.com

The Tractor Company
http://www.tractorco.com

Tractor Supply Company - TSC
http://www.tractorsupplyco.com

Tractor Web
http://www.tractorweb.com

United Breeders Inc.
http://www.ubi.com/

USAg Recycling
http://www.usagrecycling.com/

Walt's Tractor Parts, LLC
http://www.waltstractors.com/

The Waterford Company
http://www.waterfordcorp.com/

WATT Publishing Company
http://www.wattnet.com

WestCo Fertilizer
http://www.westcoag.com

Westfalia Dairy Systems
http://www.westfaliadairy.com

Willsie Equipment Sales Inc.
http://www.willsie.com
Manufacturers and distributors of fruit and vegetable equipment.

Yahoo's Agricultural Company List
http://www.yahoo.com/Business_and_Economy/Companies/Agriculture/

Yesterday's Tractors
http://www.olympus.net/biz/pratt/pratt.htm

Yocom-McColl Laboratories
http://www.ymccoll.com

Zeneca Ag Products
http://www.zenecaagproducts.com

AGRICULTURAL MAGAZINES ONLINE

Agri-Alternatives
http://www.agrialt.com

Agricultural-Related Electronic Magazines
http://www.interaccess.com/consulting/agpubs.html
A list of links to ag magazines and resources.

AgInnovator
http://www.agriculture.com/technology/index.html

ANGUS - The Magazine
http://www.cwo.com/~jdainc/angus/angidx.html

Beef Today
http://www.beeftoday.com

California Cattleman
http://www.cwo.com/~jdainc/ccm/ccmidx.html

CPM Magazine Online
http://crop-net.com

Dairy Today
http://www.dairytoday.com

Farm Journal Today
http://www.farmjournal.com

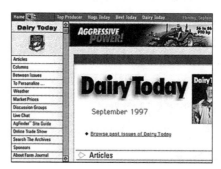

High Plains Journal
http://www.hpj.com

Hogs Today
http://www.hogstoday.com

The Home Farm
http://www.homefarm.com

Limousin World Magazine
http://www.agrione.com/LimousinWorld/

List of Electronic Magazines
http://www.agpr.com/consulting/zines.html

Progressive Farmer Online
http://pathfinder.com/PF

Ranch & Rural Living Magazine
http://biz3.iadfw.net/ranchmag

Successful Farming Magazine
http://www.agriculture.com/contents/sfonline/index.html

Top Producer
http://www.toproducer.com

UT Agriculture
http://funnelweb.utcc.utk.edu/~utia/dev/magazine.html

The Western Producer
http://www.producer.com

AGRICULTURAL ORGANIZATIONS ONLINE

Agricultural Professional Associations
http://www.agri-associations.org

Agricultural Retailers Association
http://www.agretailerassn.org

Agway, Inc.
http://www.agway.com

American Angus Association
http://www.angus.org

American Corn Growers Association
http://www.acga.org

American Egg Board
http://www.aeb.org/

American Farm Bureau-Voice of Agriculture
http://www.fb.com/home.shtml

American Farmland Trust
http://www.farmland.org

American Feed Industry Association
http://www.afia.org/

American Forage and Grassland Council
http://www.forages.css.orst.edu/Organizations/Forage/AFGC/index.html

American Gelbvieh Association
http://www.gelbvieh.org/~aga

American Jersey Cattle Association
http://www.usjersey.com

American Poultry Association
http://www.radiopark.com/apa.html

American Simmental Association
http://www.simmgene.com/asawebpg.htm

American Society of Farm Managers and Rural Appraisers
http://www.agri-associations.org/asfmra/index.html

American Soybean Association
http://www.oilseeds.org/asa

The American Texel Sheep Association
http://www.texel.org/

Association of Consulting Foresters of America, Inc.
http://www.acf-foresters.com/

Cotton Incorporated
http://www.cottoninc.com

Future Farmers of America
http://www.agriculture.com/contents/FFA/

Holstein Association
http://www.holsteinusa.com

International Sugar Association
http://www.sugarinfo.co.uk/ISO.html/

Kentucky Cattlemen's Association
http://www.kycattle.org

Michigan Farm Bureau
http://www.fb.com/mifb/

Minnesota Farmers Union
http://www.mfu.org

National Association of Wheat Growers
http://www.wheatworld.org

National Cattleman's Association
http://www.ncanet.org

National Cattlemen's Beef Association
http://www.beef.org

National Christmas Tree Association
http://www.christree.org

National Contract Poultry Growers
http://www.web-span.com/pga

National Corn Growers Association Home Page
http://www.ncga.com

National Cotton Council
http://www.cotton.org

National Council of Farmer Cooperatives
http://www.access.digex.net/~ncfc/

National Farmers Organization
http://nfo.org

National Farmers Union
http://www.nfu.or

National Mastitis Council
http://www.nmconline.org/home.htm

The National Pest Control Association
http://www.nationalpest.org

National Pork Producers Council
http://www.nppc.org

Ohio Pork Producers Council
http://www.ohiopork.org

Ontario Corn Producers' Association
http://www.ontariocorn.org

Pennsylvania Farm Bureau
http://www.fb.com/pafb/

Pennsylvania Society for Biomedical Research
http://www.psbr.org/

The Refined Sugar Association
http://www.sugarinfo.co.uk/RSA/RSA.html

U.S. Feed Grains Council
http://www.grains.org

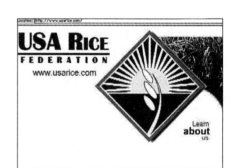

U.S. Wheat Associates
http://ianrwww.unl.edu/ianr/agronomy/wheatlab/uswt.htm

hot site **The USA Rice Federation**
http://www.usarice.com/

AGRICULTURAL RELATED NEWSGROUPS

news:ab.gov.agriculture.barley
news:alt.agriculture.fruit
news:alt.agriculture.misc
news:alt.agriculture.ratite
news:alt.sustainable.agriculture

news:clari.biz.industry.agriculture
news:clari.biz.industry.agriculture.pr
news:clari.biz.industry.agriculture.releases
news:clari.biz.market.commodities.agricultural
news:clari.web.biz.industry.agriculture
news:clari.web.biz.industry.agriculture.releases
news:clari.web.biz.market.commodities.agricultural
news:gov.us.topic.agri.farms
news:gov.us.topic.agri. food
news:gov.us.topic.agri.misc
news:gov.us.topic.agri.statistics
news:misc.rural
news:sci.agriculture
news:sci.agriculture.beekeeping
news:sci.agriculture.fruit
news:sci.agriculture.poultry
news:sci.agriculture.ratites

AGRICULTURAL SOFTWARE

AgDecisions Software
http://www.agdec.com

Agri-Logic
http://www.agrilogic.com

AgriSoft
http://www.agrisoft.com

Agro Corporation
http://www.agroco.com

Farm Home Offices
http://www.netins.net/showcase/fho

Farm Works Software
http://www.farmworks.com

Farmers Software Association
http://www.farmsoft.com

Finpack
http://www.cffm.umn.edu

Great Plains Charting Software
http://www.greatplainsusa.com/gpa.htm

Harvest Computer Systems
http://www.fmsharvest.com

Red Wing Business Systems
http://www.redwingsoftware.com

Texas A&M Extension Service Software
http://leviathan.tamu.edu

ALTERNATIVE AGRICULTURE SITES

Alternative Agriculture Publications
gopher://ndsuext.nodak.edu:70/11/agnr/alt-ag

Alternative Crops
http://hammock.ifas.ufl.edu/txt/fairs/19313

Alternative Farming Systems Information Center
http://www.nal.usda.gov/afsic/

APIS - Apicultural Information and Issues
http://www.ifas.ufl.edu/~mts/apishtm/apis.htm

AquaNIC - Aquaculture Network Information Center
http://www.ansc.purdue.edu/aquanic/

Beekeeping Home Page
http://weber.u.washington.edu/~jlks/bee.html

Crickhollow Organic Farm
http://www.telapex.com/~farmer1

Cyberfarm
http://w3.ag.uiuc.edu:80/infoag/cyberfarm/
This site is all about cyberfarming — information and information technologies that help in decision making on the farm.

Don't Panic, Eat Organic
http://www.rain.org/~sals/my.html
An organic grower's home page.

The Electronic Precision Farming Institute
http://pasture.ecn.purdue.edu/~mmorgan/PFI/PFI.html
The electronic Precision Farming Institute was developed as a WWW site for display of the latest information available on precision farming technologies.

International Bee Research Association
http://www.cardiff.ac.uk/ibra/index.html

The Kerr Center for Sustainable Agriculture
http://www.kerrcenter.com

Small Scale Farm Alternatives
gopher://psupena.psu.edu:70/1%24m%2010199248

Sustainable Agriculture and Rural Development
http://iisd1.iisd.ca/ic/info/ss9507.htm

Sustainable Agriculture Network
http://www.ces.ncsu.edu/san/

University of California Sustainable Agricultural Research
http://www.sarep.ucdavis.edu
Includes useful information on sustainable ag, including a database of cover crop information and a searchable archive of newsletter articles.

CROP RESOURCES

GENERAL CROP LINKS

Alberta Agriculture Information Sites for Crops
http://www.agric.gov.ab.ca

American Horticultural Society
http://emall.com/ahs/ahs.html

Animal and Plant Trade and Import/Export Information
http://www.aphis.usda.gov/oa/imexdir.html
USDA's Animal and Plant Health Inspection Service (APHIS) is responsible for enforcing regulations governing the import and export of plants and animals and certain agricultural products.

Australian National Botanic Gardens
http://155.187.10.12/anbg.html

Biological Control: A Guide to Natural Enemies in North America
http://www.nysaes.cornell.edu/ent/biocontrol/
This site contains photographs, descriptions of life cycles and habits, and other useful information about each natural enemy.

California Crop Improvement Association
http://www.vgl.ucdavis.edu/~ccia

California Food and Agricultural Code
http://waffle.nal.usda.gov/agdb/ca_ag_lw.html

Common Plant Diseases
http://www.ces.ncsu.edu/depts/pp/notes/
This site has links to many papers on field crop, fruit, and vegetable diseases.

Crop Information from Missouri
http://www.ext.missouri.edu/publications/xplor/agguides/crops/index.htm
Contains links to descriptions of useful reference tables you can order.

Crop Management Information from Virginia Tech
http://www.ext.vt.edu/

Crop Publications from Colorado State Cooperative Extension
http://www.colostate.edu/Depts/CoopExt/PUBS/CROPS/pubcrop.html

Daily Cash Grain Reports from the University of Kentucky
gopher://shelley.ca.uky.edu:70/11/agmkts/market_wire/grain
This site contains state by state market reports.

▼ *Part II*

Internet Addresses

Daily Future Grain Reports from the University of Kentucky
gopher://shelley.ca.uky.edu:70/00/.agwx/usr/markets/usda/MSGR711
Contains price information about wheat, corn, oats, soybeans, and soybean by-products.

Ecostat, Inc.
http://lakeland.tsolv.com/~ecostat/
This site has up-to-date information on Integrated Pest Management (IPM).

Field Crop Advisory Team CAT Reports
http://www.msue.msu.edu/msue/imp/modc2/masterc2.html

Forage Information System
http://web.css.orst.edu/
Provides links to worldwide forage information sites.

Garden Gate
http://www.prairienet.org/garden-gate/
Includes software, lists of links, gardening newsletters, and more.

Garden Net
http://trine.com/GardenNet/
Contains information about gardens on-line, catalog centers, GardenNet Magazine, garden associations, events, books, and Internet resources.

Global Agribusiness Information Network
http://www.milcom.com/fintrac/
Information about vegetable marketing, production, post-harvest handling, technology, trade shows, and other links are located at this site.

GrainGenes
gopher://greengenes.cit.cornell.edu:70/1
This gopher site includes information about pedigrees, cultivars, variety protection, genetics, software, and mapping data, and it also links to other genomes.

Hortsense from WSU
http://www.cahe.wsu.edu/~lenora/
This project was designed to provide the user with both cultural and chemical remedies for the most common yard and garden plant problems occurring in the Pacific Northwest.

151

The Master Gardener Information
http://leviathan.tamu.edu:70/1s/mg
This site contains information on fruits and nuts, flowering plants (annual and perennial), ornamental trees and shrubs, turf grasses, and vegetables. It is also searchable by single or multiple keywords.

Missouri Botanic Gardens Home Page
http://www.mobot.org/welcome.html
This site contains links to the Missouri Historical Society, Center for Plant Conservation, American Association of Botanical Gardens and Arboreta, and MBGnet for kids.

National Food and Agricultural Policy Project
http://www.eas.asu.edu/~nfapp/nfapp.htm
Contains links, policy analysis, economic analysis, market analysis, and statistics of the fruit and vegetable industry.

National Integrated Pest Management Information System at NCSU
http://ipmwww.ncsu.edu/
Provides links of interest to the urban producer, those trying to maintain a garden, and the mass crop producer.

National Pest Management Materials Database at Purdue
http://info.aes.purdue.edu/AgResearch/agreswww.html
Information includes upcoming seminars, agricultural news, and a report database.

National Plant Data Center
http://trident.ftc.nrcs.usda.gov/npdc/10links.html
Consists of an extensive list of plant related sites.

NewCROP from Purdue
http://www.hort.purdue.edu/newcrop/home
This site allows you to search an indexed database for crop information, view the NewCropNEWS newsletters, and find out about announcements and upcoming events.

Pest Management Information from Clemson
gopher://entoinfo.clemson.edu
This site features information about harmful non-indigenous species in the United States, integrated pest management, IPMnet, pest alerts, pesticide and sustainable agriculture, Pesticide Action Network North America (PANNA), and pesticide registration, news, and training.

Plant Genome WWW Page
http://probe.nalusda.gov:8000/index.html
Allows access to many databases such at the Plant Genome, Livestock Animal Genome, and Biological Genome.

Plant Pest and Disease Information
gopher://hal.aphis.usda.gov:70/11/AI.d/PPDI.d
This site contains fact sheets, reports, and other information about plant pests and diseases that threaten American agriculture.

South Dakota State University Plant Science Department
http://www.sdstate.edu/~wpls/http/pscihome.html
Contains departmental information, as well as links to Plant Science Department seminars, proceedings from the Central Alfalfa Improvement Conference, forage crops information, and winter wheat variety trial results (GrainGenes Gopher).

Sustainable Agriculture Information from the University of California
http://www.sarep.ucdavis.edu
This paper is an effort to identify the ideas, practices and policies that constitute our concept of sustainable agriculture.

Sustainable Practices for Vegetable Production in the South
http://www2.ncsu.edu/ncsu/cals/sustainable/peet/

USDA Crop Reports
gopher://gopher.nalusda.gov:70/11/ag_pubs/frmbl95/background

Vegetable Crop and Horticulture Information
gopher://cesgopher.ag.uiuc.edu:70/11/Crops-Horticulture
Contains seasonal crop information, as well as current events and newsletters.

Vegetable Crops from the University of Florida
http://hammock.ifas.ufl.edu/text/aa/39584.html
Contains information on vegetable crops, service programs, and vegetable crop publications.

Vegetable Gardening Handbook
http://hammock.ifas.ufl.edu/text/vh/19996.html
Contains information on general gardening, individual vegetables, irrigation, pests and control, propagation, soils/fertilizers, and specialized gardening.

Virtual Plant and Pest Diagnostic Laboratory
http://www.btny.purdue.edu/ppdl/default.html
Contains information concerning the laboratory, the latest pest activity and disease updates, submitting samples, newsletters, and a calendar of events.

Weed Control in Field Crops and Pasture Grasses
http://hammock.ifas.ufl.edu/text/wg/39635.html
Contains information on weed control in corn, cotton, pastures and range land, peanuts, rice, small grains, sorghum, soybeans, Florida sugarcane, and tobacco. Also has information on post emergence directed sprays for agronomic crops.

APPLES

All About Apples
http://www.dole5aday.com/about/apple/apple1.html

Apple Newsletters - North Carolina Coop. Ext.
http://henderson.ces.state.nc.us/newsletters/apple/

Apple Scab: Fruit infection
http://ppathw3.cals.cornell.edu/profiles/applescab/pa2203t6.html
Provides information about the Apple Scab fruit infection (including Apple Scab symptoms).

Apples & Such
http://www.kokodir.com/Apples.html

Canyon Park Orchard
http://apples-n-garlic.com/cpo.html
The Canyon Park Orchard picks, packs and sells over 40 varieties of apples during harvest.

Cider Space
http://www.teleport.com/~incider
This site is all about cider.

David Berkley Collection
http://dberkley.com
Order hand packaged, seasonal fruit online through this site.

Lifetime Main Ingredient
http://www.lifetimetv.com/healthtimes/main_ingredient/crwal.htm
The Recipe Database allows users to select recipes consisting of their own ingredients and favorite foods.

Northwestern Fruit & Produce Co.
http://www.wolfenet.com/~nwfruit/
This company grows, packs and sells Washington State apples and pears (as well as other fruits) worldwide.

University of Maine Cooperative Extension Apple IPM Program
http://pmo.umext.maine.edu/apple/applpage.htm

USDA Tree Fruit Research Laboratory
http://www.tfrl.ars.usda.gov/

The Virtual Orchard
http://orchard.uvm.edu/

Viruses that Infect Pome Fruit
http://www.tricity.wsu.edu/~gmink/nrspvirp.html
This page contains links to information about apple diseases.

Washington Apple Commission
http://www.treefruit.com/wac/

BERRIES

The Cranberry Home Page
http://www.scs.carleton.ca/~palepu/cranberry.html

New England Country Cupboard Cranberries
http://www.xmission.com/~arts/necc/neccmain.html

Northwest Berry & Grape InfoNet
http://www.orst.edu/dept/infonet
Provide information and communications resources for the berry and grape industries of the Pacific Northwest.

Oregon Berry Commission
http://www.peak.org/~berrywrk/

Strawberry Fieldworks
http://www.calstrawberry.com

Whistling Wings Berry Farm
http://www.biddeford.com/wwf/

CANOLA

Canola - An Alternative Crop in Indiana
http://hermes.ecn.purdue.edu:8001/http_dir/acad/agr/extn/agr/acspub/acsonline/AY-272
Contains information on field selection, fall preplant fertilization, variety selection, seedbed preparation, planting date, cost of production and economics, etc.

The Canola Connection
http://www.canola-council.org

Canola Harvest Management
http://www.agric.gov.ab.ca/crops/canola/harvest1.html
Contains information about harvesting canola, windrowing canola, chemical desiccation and pod sealant combine harvesting operations, etc.

Canola Index
http://www.agric.gov.ab.ca/crops/canola/index.html

Canola Information from North Dakota State University
http://ndsuext.nodak.edu/extnews/procrop/rps/

Canola Information from the University of California
http://pubweb.ucdavis.edu/documents/coopext/canola.htm
Contains descriptions and market and cultural information about Canola Brassica napus (Argentine type) and Brassica campestris (Polish type).

Canola Information Site: Greenland Corporation
http://canola.com/gr_corp/index.html

Canola Production from North Dakota State University Extension Service
http://www.ext.nodak.edu/extpubs/plantsci/crops/a686w.htm

Canola and Rapeseed as Enhancers of Soil Nutrient Availability and Crop Productivity in Cereal Rotations
http://pprc.pnl.gov/pprc/rpd/fedfund/usda/sare_w/canola.html

Chinese and Canadian Researchers' Partnership on Canola
http://www.idrc.ca/corp/ecanola.html

Controlling Canola Diseases in Direct Seeding Systems
http://www.agric.gov.ab.ca/agdex/500/1900001.html

Storage of Canola
http://www.agric.gov.ab.ca/crops/canola/storage1.html

Ultrabred Seed
http://fis.awp.com/Ultrabred/

CITRUS FRUITS

All About Citrus
http://www.dole5aday.com/about/citrus/citrus2.html

Citrus Fruits
http://hammock.ifas.ufl.edu/txt/fairs/ae/1933.html

Citrus Publications
http://pom44.ucdavis.edu/citruspb.html

Ferris' Valley Groves
http://www.valleygroves.com/valleygroves/

Florida Agriculture Facts - Citrus
http://gnv.ifas.ufl.edu/WWW/FLAG/FLCITRUS.HTML

Florida Citrus News & Information
http://www.floridajuice.com/floridacitrus/aninfo/index.htm

Indian River Citrus
http://indian-river.fl.us/citrus/index.html

Oranges and Soft Citrus
http://www.cosmosnet.net/azias/cyprus/frouta4.html

Pleasant Valley Ranch
http://www.netpub.com/oranges/

Pollination of Citrus by Honey Bees
http://hammock.ifas.ufl.edu/txt/fairs/aa/1216.html

Reducing Insecticide Use and Energy Costs in Citrus Pest Management
http://www.energy.ca.gov/energy/agprogram/AEAP-TEXT/PUBS/CITRUS1.HTM

Seald-Sweet Citrus Growers
http://www.sealdsweet.com/

Whidden Citrus
http://www.cyberauto.com/whidden/

COFFEE

American Coffee & Espresso
http://www.bid.com/bid/cima/

Cafe Forum
http://www.bena.com/lucidcafe/cafebystate/pennforum.html
This searchable site includes a forum, gallery, and library.

Cafe Mam Organically Grown Coffee
http://mmink.com/mmink/dossiers/cafemam.html

Caffeine Archive
http://www.austinlinks.com/General/caffeine.html

Choice Coffee - The Iced Coffee Experience
http://www.worldwidemart.com/choice/iced.html
Describes how to prepare iced coffee.

The Coffee and Caffeine FAQ
http://daisy.uwaterloo.ca/~alopez-o/caffaq.html

Coffee and Java
http://www.coffee.thelinks.com/

Coffee Islands of Hawaii
http://www.pete.com/coffee/
Features Kona Coffee, flowers, snacks and other gift items from the Hawaiian Islands.

Coffee on the Web
http://www.tiac.net/users/ckummer/coflist.htm

Coffee Recipes
http://www.javapalace.com/main/recipe.html

Coffee Source Home Page
http://www.coffee.co.cr

Coffee, Tea, & Spice
http://www.best.com/~blholmes/coffeeandtea/

Coffee Time!
http://www.yi.com/home/HammondsNancy/coffea.html

Fredricksburg Coffee Company
http://www.cains.com/coffee/
Gourmet coffee sales.

The Gourmet Coffee Club
http://www.cyber-biz.net/coffee/

HandiLinks To Coffee & Tea
http://www.ahandyguide.com/cat1/c/c75.htm

Limache Coffees on Expressbooks.com
http://expressbooks.com/info8.html
Combines books and coffee on the Internet. Features straights, blends, flavored coffees and Limache's Own Blend Books and Coffee Estate.

Java Link: Ye Olde Internet Cafe & Coffee Resource Page
http://astro.ocis.temple.edu/~ghinkle/java.html

Orleans Coffee Exchange
http://www.orleanscoffee.com/
Offers gourmet coffee from the heart of the French Quarter — over 100 varieties of origin specific coffee, organic coffee, estate coffee, blends, etc.

Sally's Place: Coffee
http://www.bpe.com/drinks/coffee/index.html

Seattle's Best Coffee
http://www.halcyon.com/sbc/sbc.html

The World of Coffee
http://www.nwlink.com/~donclark/java/world.html

CORN

American Corn Growers Association
http://www.acga.org

Corn Growers Guidebook from Purdue University
http://www.agry.purdue.edu/agronomy/ext/corn/cornguid.htm
This site contains links to corn information, including current newsletters, news releases, corn management information, and other corn related WWW sites.

Corn Information from Virginia Tech
http://www.ext.vt.edu:4040/eis/owa/docdb.getcat?cat=ir-cg-gr-co

Corn Information from the University of Missouri
http://www.ext.missouri.edu/publications/xplor/agguides/crops/#Corn
Lists available extension publications for purchase, with some documents available online.

Herbicide Injury Symptoms on Corn and Soybeans
http://www.btny.purdue.edu/Extension/Weeds/Herbinj/InjuryMOA1.html

Jolly Time Pop Corn
http://www.jollytime.com/

▼ *Part II*

Maize Genome Database
http://www.agron.missouri.edu/

National Corn Growers Association Home Page
http://www.ncga.com/

COTTON

Agriculture Marketing Service, Cotton Division
http://www.ams.usda.gov/cotton/
Contains cotton information, such as cotton programs, services, and resources, publications, and employment opportunities.

Cotton Incorporated
http://www.cottoninc.com

Cotton Information from Missouri
http://www.ext.missouri.edu/publications/xplor/agguides/crops/#Cotton
Lists available cotton reports for purchase from the extension service, with some documents available online.

Cotton Marketing and Management
http://ag.arizona.edu/AREC/cotton/Cotton-Index2.html
Contains archive of newsletters and market information. Requires Adobe Acrobat Reader to view newsletters.

Cotton Marketing Weekly from Mississippi State
http://www.ces.msstate.edu/~oac/cotton/

Cotton Sites from CottonDB Homepage
http://algodon.tamu.edu/othercot.html

CottonDB
http://algodon.tamu.edu

International Cotton Advisory Committee
http://www.icac.org/icac/english/main.html

King Cotton Magazine
http://cotton.net/

National Cotton Council of America
http://solstice.crest.org/social/eerg/ncca.html

New York Cotton Exchange
http://www.nyce.com/

HAY AND PASTURE

Alfalfa International Fact Sheet
http://www.forages.css.orst.edu/Topics/Species/Legumes/Alfalfa/International_Fact_Sheet.html
Description, area of adaptation, primary use, grazing, cultivars, fertility and ph requirements, germination and seedling development, cutting management, hay quality, diseases, and insect management of alfalfa are available here.

Field and Forage Crops from the Ag Agent Handbook
http://hammock.ifas.ufl.edu/text/aa/39582.html
Contains information on species and variety selection, forage grasses, forage and cover crop legumes, seedbed preparation, inoculation of legumes, planting dates, rates and methods, liming, lowering soil pH, fertilization, etc.

Forage Information from Virginia Tech
http://www.ext.vt.edu:4040/eis/owa/docdb.getcat?cat=ir-cg-fo
This site contains articles about the economics of one type of forage compared with another, the type of hay preferred for certain animals, storing hay, etc.

Forage Information Systems
http://web.css.orst.edu

Forage Species
http://www.cas.psu.edu/docs/casdept/agronomy/forage/docs/species/species.html
Contains information about forage species' characteristics and adaptation, establishment, hay or grazing harvest management, fertility, and pests.

Hay and Silage Harvest and Preservation
http://www.cas.psu.edu/docs/casdept/agronomy/forage/docs/haysilage/haysilage.html
Contains information about fertility, harvest management, and preservation of hay and silage.

Kansas State University Range Research
http://spuds.agron.ksu.edu/

Pasture Information
http://www.cas.psu.edu/docs/casdept/agronomy/forage/docs/pastures/pastures.html
Contains information on how to extend the life of the pasture's plants, what kind of pasture to keep for certain animals, grazing management, and economics.

Perennial Rye Grass Information
http://www.forages.css.orst.edu/Topics/Species/Grasses/Perennial_rye-grass/index.html
Contains information on perennial rye grass, including description, cultivars, adaptation, uses, establishment, fertility, management, vendors, and organizations.

SDSU Forage Information
http://www.sdstate.edu/~wpls/http/forage1.html
Includes information about the Southeast South Dakota Research Farm, SDSU Agronomy Farm, Northeast South Dakota Crops Research Farm, Central South Dakota Crops Research Farm, and contains a list of the tested alfalfa cultivars.

Soil Fertility for Forage Crops
http://www.cas.psu.edu/docs/casdept/agronomy/forage/docs/fertility/fertility.html
Contains recommendations for pre-establishment, at-establishment, and maintenance of soil fertility for forage crops.

Texas A&M Ranching System Group Home Page
http://ranch.tamu.edu/rsg/
The Ranching Systems Group was formed to develop decision support systems (DSS) for management of grazing lands. Their site contains links to product information and general ranch information.

INDUSTRIAL CROPS

The California Cut Flower Commission
http://www.flora-source.com/ccfc/index.htm

The Industrial Hemp Information Network: News
http://hemptech.com/hnews.html

National Christmas Tree Association
http://www.christree.org

OTHER FRUITS

California Avocado Commission
http://www.avoinfo.com

California Cherry Advisory Board
http://www.califcherry.com

David Berkley Collection
http://dberkley.com
Order hand packaged, seasonal fruit online.

Fruit Crops
http://hammock.ifas.ufl.edu/txt/fairs/19943
Contains information about varieties, pollination, propagation, planting, cultivation, fertilization, irrigation, pruning, harvesting, and storage.

Fruit Facts
http://www.crfg.org/pubs/frtfacts.html

The Master Grape Page
http://www.geocities.com/NapaValley/2680/
This is a page about grapes. Contains some fun links and some serious links.

Met West Agribusiness
http://www.sunmet.com/

Scaffold's Fruit Journal
http://www.nysaes.cornell.edu/ent/scafolds/

Summertime Fresh Fruits From Farm Direct
http://www.farmdirect.com/vendors/summertime/

Washington Fruit Commission
http://www.nwcherries.com/

PEANUTS

The NESPAL Peanut Literature Database
http://nespal.cpes.peachnet.edu/pnutdb/pnutlib.html

The Snickers(tm) Peanut Gallery
http://www.panam.snickers.com/Snickers/Facts/facts.html
This site is just for fun :-)

Virginia Carolina Peanuts
http://aboutpeanuts.com

Yahoo's Peanut Information
http://www.yahoo.com/Business_and_Economy/Companies/Food/Produce/Nuts/Peanuts/

RICE

About Rice Genes
http://probe.nalusda.gov:8000/plant/aboutricegenes.html

AGIS: Databases: Rice Genes
http://probe.nalusda.gov:8300/cgi-bin/dbrun/ricegenes?find+Map

Broussard Rice Mill
http://www.inti.net/brmi.htm

Climate Change and Rice
http://ourworld.compuserve.com/homepages/rbmatthews/rbm_cc1.htm

Crop Estimates for Rice
http://usda.mannlib.cornell.edu:70/data-sets/crops/95501

Crop Production - Acreage for Rice
http://usda.mannlib.cornell.edu:70/reports/nassr/field/pcp-bba

Crop Production - Prospective Plantings
http://usda.mannlib.cornell.edu:70/reports/nassr/field/pcp-bbp

Crop Progress and Condition
http://usda.mannlib.cornell.edu:70/data-sets/crops/93115

Crop Values for Rice
http://usda.mannlib.cornell.edu:70/reports/nassr/price/zcv-bb

Grey Owl Foods - Wild Rice Project
http://www.greyowlfoods.com/siap.html

GRIN - Germplasm Resources Information Network
http://www.ars-grin.gov/npgs/

Kagiwiosa Manomin Wild Rice
http://dryden.lakeheadu.ca/~manomin/home.html

Lundberg Family Farms
http://www.lundberg.com

Overview of Florida Rice
http://hammock.ifas.ufl.edu/txt/fairs/15619

The Rice Company
http://www.riceco.com/

Rice Genes
http://probe.nalusda.gov:8300/cgi-bin/browse/ricegenes

Rice Genetics Newsletters, Journals and Other Publications
http://probe.nalusda.gov:8000/otherdocs/

Rice Genetics Project
http://agronomy.ucdavis.edu/Mackill/Homepage.html

Rice Genome Program (Japan)
http://www.staff.or.jp/

Rice Information from NewCrop
http://www.hort.purdue.edu/newcrop/crops/rice
Contains information on history, varieties, rice culture, milling and uses of rice.

Rice Policy
http://ianrwww.unl.edu/farmbill/rice.htm

Rice Reports
http://ssu.agri.missouri.edu/ssu/fapri/reports/195/execsum/text/foodg.htm

Rice Stocks
http://usda.mannlib.cornell.edu:70/reports/nassr/field/prs-bb

Rice Stocks: Final Estimates
http://usda.mannlib.cornell.edu:70/data-sets/crops/94898

Rice Yearbook (RCS)
http://usda.mannlib.cornell.edu:70/reports/erssor/field/rcs-bby

TAEX Rice Information
http://leviathan.tamu.edu:70/7c/.cache?rice

U.S. Rice Federation
http://www.usarice.com

U.S. Rice Industry Basebook Data
http://usda.mannlib.cornell.edu:70/data-sets/crops/94020

World Rice
http://ssu.agri/missouri.edu/ssu/fapri/reports/195/wtrade/text/wrice.htm

SMALL GRAINS

Barley Information from NewCrop
http://www.hort.purdue.edu/newcrop/crops/barley
This site describes the history, characteristics, cultivated variety groups, and uses of barley.

Combine Talk Shows
http://www.harvesting.com/combine/directory.htm
This site allows you to ask and answer combine/grain harvesting related questions, much like an online BBS.

The Kansas Wheathearts
http://www.hpj.com/whearts.htm
This is an educational site about wheat and grains aimed at kids and non-farmers, but should appeal to all.

Oats Information from NewCrop
http://www.hort.purdue.edu/newcrop/crops/oats
This site contains general information, botanical classification, and uses of oats.

Rye Information from NewCrop
http://newcrop.hort.purdue.edu/newcrop/crops/rye
This site contains general information about rye.

Wheat Foods Council
http://www.wheatfoods.org

Wheat Information from NewCrop
http://www.hort.purdue.edu/newcrop/crops/wheat
Contains general information, botanical classification, and uses of wheat.

Winter Wheat Production Information
http://www.usask.ca/agriculture/cropsci/winter_wheat/

SORGHUM

AGIS Sorghum Database (Experimental Site)
http://probe.nalusda.gov:8300/cgi-bin/browse/sorghumdb

Cover Crop Database: Complete Crop Summary of Sorghum & Sudangrass
http://www.sarep.ucdavis.edu/sarep/ccrop/crops/crop36.htm

Grain Sorghum: Fertilization and Pest Control
http://www.aac.msstate.edu/pubs/is1225.htm

The Kansas Grain Sorghum Producers Association
http://www.kanza.net/%7Esorghum/

Micnhimer's Sorghum
http://www.neosoft.com/~sorghum/tlclocal.htm

Sorghum Dictionary
http://www.met.unimelb.edu.au/Porcher/Sorghum.html

Sorghum Disease Index
http://cygnus.tamu.edu/Texlab/Grains/Sorghum/sortop.html

Sorghum Research Data
http://www.agriculture.com/contents/imc/research/SorgRS.HTML

Sorghum Research Project
http://sun.ars-grin.gov/ars/SoAtlantic/Mayaguez/sorghum.html

Statistical Analysis Of 8000 B.P. Sorghum Remains From The Nabta Playa
http://www.nal.usda.gov/ttic/tektran/data/000007/29/0000072992.html

The USDA Sorghum Research Project
http://www.ars-grin.gov/ars/SoAtlantic/Mayaguez/sorghum.html

SOYBEANS

American Dry Bean Board
http://www.prairieweb.com/bean

American Soybean Association Homepage
http://www.oilseeds.org/asa/welcome.htm

National Soybean Research Laboratory
http://www.ag.uiuc.edu/~nsrl/nsrlpage.html
This site has links to information on the National Soybean Research Laboratory, as well as links to other sources of soybean information.

North Carolina Soybean Information
http://ipmwww.ncsu.edu/soybeans/soybean_contents.html
Contains information concerning insects, diseases, weeds, nematodes, production, pesticides and wildlife.

Soy Stats
http://www.ag.uiuc.edu/~asa/soystat/soystat.html
Contains soybean price and value information, uses, trade, and links to other sites of interest.

Soyatech, Inc.
http://www.soyatech.com/Soyatech.html

StratSoy
http://www.ag.uiuc.edu/~stratsoy/stratsoy.html
StratSoy is a communication and information system for the U.S. soybean industry. This site contains information about soybean organizations, resources, and research database; and you can ask an expert questions.

United Soybean Board
http://www.ag.uiuc.edu/~usb/

Yahoo's Soybean Production Information
http://www.yahoo.com/Science/Agriculture/Crops_and_Commodities/Soybeans

TOBACCO

Agriculture Marketing Service - Tobacco Division
http://www.ams.usda.gov/tob/
Links to information regarding tobacco programs, services and resources, standardization and review, and market information and program analysis.

Tobacco & Health Research Institute
http://www.uky.edu/~thri/homeweb.html

TUBERS

Center for Aquatic Plants
http://aquat1.ifas.ufl.edu/pick14.html

Giant Tubers
http://mail.coos.or.us/~bishop/tubers.htm

Idaho Potato Exposition
http://www.sisna.com/Idaho_Potato_Expo/

Maine Potato Board
http://www.mainerec.com/mepotbd.html

VEGETABLES

Broccoli Production Guide for North Carolina
http://www.ces.ncsu.edu/hil/hil-5-b.html

Cornell University Fruit and Vegetable Information
gopher://gopher.cce.cornell.edu/11/cenet/submenu/fruit-veg

Engineering and Cultivation Systems
http://www.fb.u-tokai.ac.jp/pp-info/p-eng.html

Florida Fruit and Vegetable Association
http://www.ffva.com

Friends of the UCSC Farm & Garden
http://zzyx.ucsc.edu/casfs/friends.html

Gardening Archive
http://www.lysator.liu.se/garden/index.html

Home and Gardening List from Yahoo
http://www.yahoo.com/Recreation/Home_and_Garden

Home Canning Guide
http://hammock.ifas.ufl.edu/txt/fairs/31520

Horticulture Library
http://www.oznet.ksu.edu/library/HORT2/MF2030.PDF

Horticulture Solutions
http://www.ag.uiuc.edu/~robsond/solutions/horticulture/soils.html

Sustainable Practices for Vegetable Production in the South
http://www2.ncsu.edu/ncsu/cals/sustainable/peet

Texas Plant Disease Handbook
http://cygnus.tamu.edu/Texlab/tpdh.html

UC Davis Department of Vegetable Crops
http://veghome.ucdavis.edu

USDA National Plant Data Center
http://trident.ftc.nrcs.usda.gov/npdc/10links.html

USDA Vegetable Crops Research Unit
http://www.wisc.edu/hort/usdavcru

Veg Net
http://www.ag.ohio-state.edu/~vegnet

Vegetable and Row Crop Pest Management
http://pubweb.ucdavis.edu/documents/coopext/disease.htm

Vegetable Crop Irrigation
http://www.ces.ncsu.edu/depts/hort/hil/hil-33-e.html

Vegetable Crops Hotline
http://www.entm.purdue.edu/Entomology/vc_hotline/index.html

Vegetable Crops Service Program
http://hammock.ifas.ufl.edu/txt/fairs/16292

Vegetable Gardening Handbook
http://hammock.ifas.ufl.edu/text/vh/19996.html

Vegetable Seeds Index
http://www.fred.net/sports/seedinde.html

ENTOMOLOGY

Colorado State University Entomology Department
http://www.colostate.edu/Depts/Entomology/ent.html
This page contains links to upcoming events in entomology, entomology sites on the Internet, publications, and insect pictures and movies.

Michigan State University Entomology Department
http://www.ent.msu.edu/
Contains departmental information, as well as links to entomology guides and indexes, and entomology-related Internet information.

Ohio State University Entomology Department
http://iris.biosci.ohio-state.edu/osuent/home.html
Contains general departmental links and course information.

Ohio State University Insect Collection
http://iris.biosci.ohio-state.edu/inscoll.html
This page contains information on taxonomic catalogs of hymenoptera, hymenoptera newsletters on-line, USDA/APHIS/PPQ Pest Alerts, etc.

Purdue University Entomology Department
http://www.entm.purdue.edu/entomology/entmwww.html
This site contains links to extension publications and newsletters, the Center for Urban and Industrial Pest Management, the Biological Control Laboratory, and the Virtual Pest and Plant Diagnostic Laboratory.

University of Arizona Entomology Department
http://ag.arizona.edu/ENTO/entohome.html
This site has links to the Carl Hayden Honeybee Lab and the Center for Insect Science, as well as links to information about the research facilities on the main campus in Tucson and throughout the State of Arizona.

University of Kentucky Entomology Department ENTFACTS
http://www.uky.edu/Agriculture/Entomology/enthp.htm
Includes tons of entomology information, including "Search the Site" and "Ask the Expert" features.

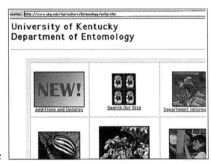

University of Massachusetts-Amherst Entomology Department
http://www.umass.edu/ent/index.html
Contains general departmental links and course information.

University of Missouri Entomology Department
http://forent.insecta.missouri.edu
Contains general departmental links and course information.

University of Nebraska Insect Ecology
http://unlvm.unl.edu/iecol.htm

FARMERS ONLINE

5T Ranch
http://gator.net/~tefertil/5t/

A Cowman's Choice - McKenney Farms
http://www.cowmans.com/10txcmck.htm

A & J Ranch Home Page
http://www.geocities.com/RodeoDrive/1371

Absolutely Divine Pecans
http://www.choicemall.com/divinepecans

Absolutely Fresh Flowers
http://www.cts.com/~flowers

Adam's Ranch Home Page
http://orchid-isle.com/business/adams

Agworks Online
http://www.agworks.com/breeders.htm

Alberda Angus Ranch
http://205.163.41.2/alberda

Bar W Home Page
http://www.jci.net/~sharonw/Ranching

Barry Bean
http://www.concentric.net/~Bbbean/

Beef Cattle Teaching Center
http://www.canr.msu.edu/ans/pbeefaci.html

Beefalo Home Page
http://www.beefalobeef.com

Beefitup's Home Page
http://www.geocities.com/Yosemite/4640/index.html

Blaine Cattle Ranch
http://www.moscow-id.com/business/bcrbeef/bcr3.htm

Blue Water Aquaculture
http://www.bwaqua.com/index2.html

Broke Again Farms, LTD.
http://ourworld.compuserve.com:80/homepages/jsorenson/

Chase Tavern Farm Alpacas
http://www.maine.com/ctalpacas/

Double J Ranch
http://home1.gte.net/doublej1/

Duane Bouse
http://www.geocities.com/Heartland/Plains/2056/

Dundee Hills Farm
http://www.europa.com/kiwikern
Kiwifruit Products.

Featherside Farm
http://www.cyborganic.net/people/feathersite/Poultry/BRKPoultryPage.html

Florinsa Farms
http://www.florinsa.com/

Flying W Farms Citrus Grove
http://www.atlantic.net/~flywfarm

Frescargo Farms
http://www.etropolis.com/snails/market.htm
Snail Farming.

Glover Cattle Company
http://www.catalog-catalog.com/texas/longhorn/catalog.html

Groveland Farm
http://members.aol.com/sheepmilk/www/grove.html

Hummer and Son Honey Farm
http://services.ciai.net/~whummer

Jim Bell
http://www.midwest.net/scribers/jbell/index.htm

John Reifsteck's Farm Home Page
http://w3.aces.uiuc.edu/InfoAg/CyberFarm/Reifsteck/

MacFarms Macadamia Nuts
http://macfarms.com/default.html

Miles Estate Herb and Berry Farm
http://www.herbs-spices-flowers.com/index.phtml?gid=875733245.19690700

Muddy H Holsteins
http://www.muddyh.com

Noah's Ark Organic Farm
http://www.rain.org/~sals/my.html

Owenlea Farm Home Page
http://www.bright.net/~fwo/

Perfect 10 Buffalo Ranch
http://www.solu.net/buffalo

Porter Ranches
http://www.porterranches.com/

Robert Nottelmann's Farm Management Page
http://www.maxinet.com/nott/Index.htm#top

Robinson Farms
http://www.sweetvidalias.com
Vidalia onion growers.

Sampson McGregor Stock Farm
http://www.compusmart.ab.ca/mcgregor/

Shepherd Bio - Agribusiness
http://www.bossnt.com/~gshep/shep.html

Shiloh Creek Farm
http://www.leanbeef.com/

Wheatina's Amber Waves Page
http://www.brigadoon.com/~blyle/

Witt's End Ranch
http://www.wolfenet.com/~wittzend
Registered quarter horses.

Yahoo's List of Farmer's Sites
http://www.yahoo.com/Science/Agriculture/Farmers/

FORESTRY

Appalachian Sustainable Forestry Home Page
http://www.uky.edu/OtherOrgs/AppalFor/high.html

Association for Temperate Agro-Forestry
http://www.missouri.edu/~afta/afta_home.html

Canadian Atlantic Forestry Centre
http://www.fcmr.forestry.ca/

Forestry Information
http://hammock.ifas.ufl.edu/text/aa/39585.html
Contains forestry information primarily related to Florida.

Forestry Today Magazine
http://www.forestry.com/forestry/

GAIA Forest Conservation Archives — Sustainable Forestry Documents
http://forests.org/forests/susforest.html

University of Florida — School of Forest Resources and Conservation Home Page
http://www.sfrc.ufl.edu/

University of Kentucky Department of Forestry
http://www.uky.edu/Agriculture/Forestry/forestry.html

USDA Forest Service
http://www.fs.fed.us

Virginia Tech's College of Forestry and Wildlife Resources
http://www.fw.vt.edu/

The World Wide Web Virtual Library: Forestry
http://www.metla.fi/info/vlib/Forestry

WSU Forestry Vegetation Simulator Gopher Site
gopher://gopher.cahe.wsu.edu

▼ *Part II*

LAND GRANT UNIVERSITIES

The following are the land grant universities, which are state partners of the Cooperative State Research, Education, and Extension Service.

ALABAMA

Alabama A & M University
http://www.aamu.edu

Cooperative Extension Program
http://saes.aamu.edu/exten.htm

School of Agriculture and Environmental Sciences
http://saes.aamu.edu

Auburn University
http://www.duc.auburn.edu

Alabama Agricultural Experiment Stations
http://www.acesag.auburn.edu/department/grain/AAES.htm

Alabama Cooperative Extension Service
http://gn.acenet.auburn.edu

College of Agriculture
http://www.ag.auburn.edu

College of Veterinary Medicine
http://www.vetmed.auburn.edu

School of Forestry
http://www.forestry.auburn.edu

School of Human Sciences
http://www.humsci.auburn.edu

Tuskegee University
http://www.tusk.edu

ALASKA

University of Alaska - Fairbanks
http://zorba.uafadm.alaska.edu

Alaska Cooperative Extension
http://zorba.uafadm.alaska.edu/coop-ext/index.html

ARIZONA

University of Arizona
http://www.arizona.edu

Arizona Cooperative Extension
http://ag.arizona.edu:/Ext/coopext.html

College of Agriculture
http://ag.arizona.edu

ARKANSAS

University of Arkansas
http://www.uark.edu

Arkansas Cooperative Extension
http://www.uaex.edu

College of Agricultural, Food and Life Sciences
http://www.uark.edu/campus/Academic/colleges/agri/index.html

University of Arkansas Pine Bluff
http://www.uapb.edu

CALIFORNIA

University of California
http://www.ucop.edu

Agricultural Personnel Management Program
http://are.berkeley.edu/APMP

College of Agricultural and Environmental Sciences
http://www.ucdavis.edu/aes.html

Division of Agriculture and Natural Resources
http://danr.ucop.edu

Livestock and Natural Resources
http://danr.ucop.edu/uccelr/uccelr.htm

School of Veterinary Medicine
http://www.vetnet.ucdavis.edu

UC's Integrated Pest Management Project
http://www.ipm.ucdavis.edu

University of California - Berkeley College of Natural Resources
http://www.cnr.berkeley.edu

University of California Cooperative Extension, North Region
http://www.ucce-north.ucdavis.edu

University of California Cooperative Extension, South Central Region
http://www.uckac.edu/danrscr

University of California - Riverside College of Natural and Agricultural Sciences
http://cnas.ucr.edu

COLORADO

Colorado State University
http://www.colostate.edu

College of Agricultural Sciences
http://www.colostate.edu/Depts/AgSci/agsci.html

College of Natural Resources
http://www.cnr.colostate.edu

College of Veterinary Medicine and Biomedical Sciences
http://www.vetmed.colostate.edu

Colorado Agricultural Experiment Station
http://www.colostate.edu/Depts/AES

Colorado State Cooperative Extension
http://www.colostate.edu/Depts/CoopExt

CONNECTICUT

University of Connecticut
http://www.uconn.edu

Agricultural Experiment Station
http://www.lib.uconn.edu/CANR/expsta/index.html

College of Agriculture and Natural Resources
http://www.lib.uconn.edu/CANR

Cooperative Extension System
http://www.lib.uconn.edu/CANR/ces/index.html

DELAWARE

Delaware State University
http://www.dsc.edu

University of Delaware
http://www.udel.edu

College of Agricultural Sciences
http://bluehen.ags.udel.edu

Delaware Cooperative Extension
http://bluehen.ags.udel.edu/deces

FLORIDA

Florida A & M University
http://www.famu.edu

University of Florida
http://www.ufl.edu

Florida Agricultural Experiment Station
http://www.ifas.ufl.edu/WWW/AGATOR/HTM/FLAES.HTM

Florida Agricultural Retrieval System
http://hammock.ifas.ufl.edu

Florida Cooperative Extension
http://www.ifas.ufl.edu/WWW/AGATOR/HTM/CES.HTM

Institute of Food and Agricultural Sciences
http://www.ifas.ufl.edu

GEORGIA

Fort Valley State University
http://www.fvsu.edu

The School of Agriculture, Home Economics and Allied Programs
http://agschool.fvsc.peachnet.edu

University of Georgia
http://www.uga.edu

Agricultural Experiment Stations
http://www.griffin.peachnet.edu/ugaexpstn.html

College of Agriculture and Environmental Sciences
http://www.uga.edu/~caes

College of Forest Resources
http://www.uga.edu/~wsfr/

College of Veterinary Medicine
http://www.vet.uga.edu

Cooperative Extension Service
http://www.ces.uga.edu

HAWAII

University of Hawaii
http://www.hawaii.edu/uhinfo.html

College of Tropical Agriculture and Human Resources
http://www.ctahr.hawaii.edu

IDAHO

University of Idaho
http://www.uidaho.edu

College of Agriculture
http://www.uidaho.edu/ag

Cooperative Extension System
http://decit.if.uidaho.edu/CoAg/distiv.html

Cooperative Extension System Idaho Falls R&E Center - District IV Office
http://decit.if.uidaho.edu/CoAg/distiv.html

Idaho Agricultural Experiment Station
http://www.uidaho.edu/ag

ILLINOIS

University of Illinois
http://www.uiuc.edu

Agricultural Experiment Station
http://www.ag.uiuc.edu/iaeshome.html

College of Agriculture, Consumer, and Environmental Sciences
http://w3.aces.uiuc.edu

Illinois Cooperative Extension
http://www.ag.uiuc.edu

INDIANA

Purdue University
http://www.purdue.edu

Academic Programs
http://www.agad.purdue.edu/academic/contents.htm

Agricultural Administration
http://www.agad.purdue.edu

Agricultural Research Programs
http://info.aes.purdue.edu/agresearch/agreswww.html

Cooperative Extension
http://hermes.ecn.purdue.edu:8001/http_dir/acad/agr/extn/extn.html

School of Consumer and Family Science
http://wombat.cfs.purdue.edu

Veterinary Medicine
http://www.vet.purdue.edu

IOWA

Iowa State University
http://www.iastate.edu

Department of Forestry
http://www.ag.iastate.edu/departments/forestry/Forestry.html

Iowa Agricultural and Home Economics Experiment Station
http://www.ag.iastate.edu/iaexp

Iowa State University College of Agriculture
http://www.ag.iastate.edu

Iowa State University College of Veterinary Medicine
http://www.vetmed.iastate.edu/vetmed.html

Iowa State University Extension
http://www.exnet.iastate.edu

KANSAS

Kansas State University
http://www.ksu.ksu.edu

College of Agriculture
http://www.oznet.ksu.edu/coa.htm

College of Human Ecology
http://www.ksu.ksu.edu/humec

College of Veterinary Medicine
http://www.vet.ksu.edu

Kansas Cooperative Extension
http://www.oznet.ksu.edu

KENTUCKY

Kentucky State University
http://www.state.ky.us/ksu/ksuhome.htm

University of Kentucky
http://www.uky.edu

College of Agriculture
http://www.ca.uky.edu

College of Human Environmental Sciences
http://www.uky.edu/HES

LOUISIANA

Louisiana State University
http://www.lsu.edu

Agricultural Center
http://www.agctr.lsu.edu/wwwac

College of Agriculture
http://www.stat.lsu.edu/coa/coa-home.html

Southern University
http://www.subr.edu

MAINE

University of Maine
http://kramer.ume.maine.edu

Department of Applied Ecology and Environmental Science
http://www.ume.maine.edu/~aes/

Maine Agricultural and Forest Experiment Station
http://www.ume.maine.edu/~nfa/mafes/welcome.htm

Sustainable Agriculture/Natural Resources
http://www.umext.maine.edu/susag.htm

University of Maine Cooperative Extension
http://www.umext.maine.edu

MARYLAND

University of Maryland
http://www.umd.edu

College of Agriculture and Natural Resources
http://www.agnr.umd.edu

Eastern Shore Rural Development Center
http://www.umes.umd.edu/dept/rudept.html

Maryland Agricultural Experiment Station
http://www.agnr.umd.edu/AES

Maryland Cooperative Extension Service
http://www.agnr.umd.edu/CES

University of Maryland Eastern Shore
http://www.umes.umd.edu

MASSACHUSETTS

University of Massachusetts
http://www.umass.edu

College of Food and Natural Resources
http://www.umass.edu/cfnr

Massachusetts Agricultural Experiment Station
http://www.umass.edu/maes

The Stockbridge School of Agriculture
http://www.umass.edu/stockbridge

University of Massachusetts Extension
http://www.umass.edu/umext

MICHIGAN

Michigan State University
http://www.msu.edu

Michigan State University Extension
http://www.msue.msu.edu/msue

MINNESOTA

University of Minnesota
http://www.umn.edu/tc

College of Agricultural, Food, and Environmental Sciences
http://beauty.agoff.umn.edu/~coafes

College of Human Ecology
http://www.che.umn.edu

College of Natural Resources
http://www.cnr.umn.edu

College of Veterinary Medicine
http://www.cvm.umn.edu

Continuing Education & Extension
http://www.cee.umn.edu

Minnesota Cooperative Extension
http://www.mes.umn.edu

Minnesota Agricultural Experiment Station
http://www.mes.umn.edu/~maes/

MISSISSIPPI

Alcorn State University
http://academic.alcorn.edu

Mississippi State University
http://www.msstate.edu

Division of Agriculture, Forestry, and Veterinary Medicine
http://www.ces.msstate.edu/division

Mississippi Agricultural and Forestry Experiment Station
http://www.aac.msstate.edu/mafes

Mississippi Cooperative Extension
http://www.ces.msstate.edu/ces.html

Southern Association of Agricultural Experiment Station Directors
http://www.msstate.edu/org/saaesd

MISSOURI

Lincoln University
http://www.lincolnu.edu

University of Missouri
http://www.system.missouri.edu

University of Missouri Extension
http://extension.missouri.edu

MONTANA

Montana State University-Bozeman
http://www.montana.edu

Montana Agricultural Experiment Station
http://www.montana.edu/~optjh/aghp.html

Montana State University Extension Service
http://www.montana.edu/~wwwcx/index.html

Montana State University School of Forestry
http://www.forestry.umt.edu

NEBRASKA

University of Nebraska-Lincoln
http://www.unl.edu

University of Nebraska Cooperative Extension
http://ianrwww.unl.edu/ianr/coopext/coopext.htm

University of Nebraska-Lincoln Institute of Agriculture & Natural Resources
http://ianrwww.unl.edu

NEVADA

University of Nevada - Reno
http://www.unr.edu/unr

College of Agriculture
http://www.unr.edu/colleges/agric/index.html

NEW HAMPSHIRE

University of New Hampshire
http://unhinfo.unh.edu

College of Life Sciences and Agriculture
http://arethusa.unh.edu/colsa/colsa.htm

University of New Hampshire Cooperative Extension
http://ceinfo.unh.edu

NEW JERSEY

Rutgers University
http://www.rutgers.edu

The Agricultural Economic Recovery and Development Initiative
http://aesop.rutgers.edu/www/aerdi

Cook College/New Jersey Agricultural Experiment Station
http://aesop.rutgers.edu

Rutgers Cooperative Extension
http://www.rce.rutgers.edu

NEW MEXICO

New Mexico State University
http://www.nmsu.edu

Agricultural Science Centers and Research Centers
http://www.cahe.nmsu.edu/cahe/aes/aesctrrsh.html

New Mexico State University Agricultural Experiment Station
http://www.cahe.nmsu.edu/cahe/aes

New Mexico State University College of Agriculture and Home Economics
http://www.cahe.nmsu.edu/cahe

New Mexico State University Cooperative Extension
http://www.cahe.nmsu.edu/cahe/ces

NEW YORK

Cornell University
http://www.cornell.edu

College of Agriculture and Life Sciences
http://www.cals.cornell.edu/cals

College of Human Ecology
http://www.human.cornell.edu

College of Veterinary Medicine
http://zoo.vet.cornell.edu

Cornell Cooperative Extension
http://www.cce.cornell.edu

Cornell University Agricultural Experiment Station - Ithaca
http://www.cals.cornell.edu/cals/OfficeResearch

New York State Agricultural Experiment Station - Geneva
http://www.nysaes.cornell.edu

State University of New York
http://www.suny.edu

College of Environmental Science and Forestry
http://www.esf.edu

NORTH CAROLINA

North Carolina A&T University
http://www.ncat.edu

College of Agriculture
http://www.ncat.edu/~soa

North Carolina A&T Cooperative Extension Program
http://www.ncat.edu/~soa/extension

North Carolina A&T University Agricultural Research
http://www.ncat.edu/~soa/agresearch

North Carolina State University
http://www.ncsu.edu

College of Forest Resources
http://www2.ncsu.edu/ncsu/forest_resources/cfr.html

College of Veterinary Medicine
http://www2.ncsu.edu/ncsu/cvm/cvmhome.html

North Carolina Cooperative Extension Service
http://www.ces.ncsu.edu

NORTH DAKOTA

North Dakota State University
http://www.ndsu.nodak.edu

Agricultural Experiment Station
http://www.ag.ndsu.nodak.edu/exphp.htm

College of Agriculture
http://www.ndsu.nodak.edu/instruct/mcclean/ag_www/agriculture.html

North Dakota Cooperative Extension
http://www.ext.nodak.edu

OHIO

Ohio State University
http://www.acs.ohio-state.edu

College of Food, Agricultural, and Environmental Sciences
http://www.ag.ohio-state.edu/~cfaes/

College of Human Ecology
http://www.osu.edu/units/hecology/index.html

College of Veterinary Medicine
http://www.vet.ohio-state.edu

Ohio Agricultural Research and Development Center
http://www.oardc.ohio-state.edu

Ohio State University Extension Service
http://www.ag.ohio-state.edu

Ohioline
http://www.ag.ohio-state.edu/~ohioline

OKLAHOMA

Langston University
http://www.lunet.edu

Oklahoma State University
http://pio.okstate.edu

Oklahoma Cooperative Extension Service
http://www.okstate.edu/OSU_Ag/oces

OREGON

Oregon State University
http://www.orst.edu

College of Agricultural Sciences
http://www.orst.edu/mc/coldep/agrsci/agrsci.htm

College of Veterinary Medicine
http://www.orst.edu/mc/coldep/vetmed.htm

Oregon Cooperative Extension
http://wwwagcomm.ads.orst.edu/AgComWebFile/extser/index.html

PENNSYLVANIA

Penn State University
http://www.psu.edu

College of Agricultural Sciences
http://www.cas.psu.edu

Penn State Cooperative Extension
http://www.cas.psu.edu/docs/COEXT/COOPEXT.HTML

RHODE ISLAND

University of Rhode Island
http://www.uri.edu

College of Resource Development
http://www.uri.edu/crd/crd_home.html

Rhode Island Cooperative Extension
http://www.edc.uri.edu

SOUTH CAROLINA

Clemson University
http://www.clemson.edu

Clemson Cooperative Extension
http://agweb.clemson.edu/exten/home.htm

Clemson University College of Agriculture, Forestry, and Life Sciences
http://agweb.clemson.edu

South Carolina State University
http://192.231.63.160/scsu/state.htm

SOUTH DAKOTA

South Dakota State University
http://www.sdstate.edu

College of Agricultural and Biological Sciences
http://www.sdstate.edu/~http/http/sdsuinfo/colleges/coll_agbio.html

South Dakota State Cooperative Extension
http://www.abs.sdstate.edu/CES

Veterinary Medicine and Science Department
http://www.vetsci.sdstate.edu

TENNESSEE

Tennessee State University
http://www.tnstate.edu

University of Tennessee - Knoxville
http://www.utk.edu

Agricultural Experiment Station
http://funnelweb.utcc.utk.edu/~taescomm/

College of Agriculture Sciences and Natural Resources
http://funnelweb.utcc.utk.edu/~casnr/

College of Veterinary Medicine
http://funnelweb.utcc.utk.edu/vet

Tennessee Agricultural Extension Service
http://funnelweb.utcc.utk.edu/~utext/

TEXAS

Texas A&M University
http://www.tamu.edu

College of Agriculture and Life Sciences
http://gallus.tamu.edu/coals/coals.html

Prairie View A&M University
http://www.pvamu.edu

Texas Agricultural Extension Service
http://leviathan.tamu.edu

Texas A&M Agricultural Experiment Station
http://taeswww.tamu.edu

Texas A&M Agriculture Program
http://agprogram.tamu.edu

UTAH

Utah State University
http://www.usu.edu

College of Family Life Extension
http://ext.usu.edu/famlife

Utah Agricultural Experiment Station
http://ext.usu.edu/agx

Utah State Extension
http://ext.usu.edu

VERMONT

University of Vermont
http://www.uvm.edu

University of Vermont Extension Service
http://ctr.uvm.edu/ext

Vermont Agricultural Experiment Station
http://ctr.uvm.edu/research

Vermont College of Agriculture and Life Sciences
http://ctr.uvm.edu/cals/CALS.HTM

VIRGINIA

Virginia Polytechnic Institute and State University
http://www.vt.edu/

Virginia Agricultural Experiment Station
http://www.vaes.vt.edu

Virginia Cooperative Extension
http://www.ext.vt.edu

Virginia State University
http://www.vsu.edu

WASHINGTON

Washington State University
http://www.wsu.edu

College of Agriculture and Home Economics
http://coopext.cahe.wsu.edu/cahe.html/cals

Washington Cooperative Extension
http://www.cahe.wsu.edu/ce.html

WEST VIRGINIA

West Virginia University
http://www.wvu.edu

College of Agriculture and Forestry
http://www.caf.wvu.edu

West Virginia Extension Service
http://www.wvu.edu/~exten

WISCONSIN

University of Wisconsin - Madison
http://www.wisc.edu

Department of Agricultural and Applied Economics
http://www.wisc.edu/aae

Wisconsin Cooperative Extension
http://www.uwex.edu/ces

WYOMING

University of Wyoming
http://www.uwyo.edu

Wyoming Agricultural Experiment Station
http://www.uwyo.edu/ag/agexpstn/exphome.htm

Wyoming College of Agriculture
http://www.uwyo.edu/ag/agadmin/ag.htm

Wyoming Cooperative Extension
http://www.uwyo.edu/ag/ces/ceshome.htm

LIVESTOCK RESOURCES

GENERAL LIVESTOCK LINKS

Agriculture Answers from Purdue and Ohio State
http://www.aes.purdue.edu/AgAnswrs/AgAnswers.html
This site provides timely agricultural problem-solving advice, strategies, and reminders to help farmers better manage their crops, their livestock, and their marketplace transactions.

Agriculture Marketing Service, Livestock and Seed Division at USDA
http://www.ams.usda.gov/
This site contains information on grading and certification, market news, standardization, purchase programs, research and promotion programs, seed activities, and international programs.

Animal and Plant Trade and Import/Export Information
http://www.aphis.usda.gov/oa/imexdir.html
USDA's Animal and Plant Health Inspection Service (APHIS) is responsible for enforcing regulations governing the import and export of plants and animals and certain agricultural products.

Animal Health Information
gopher://hal.aphis.usda.gov:70/11/AI.d/AHI.d
This site contains fact sheets, technical reports, and other information such as health monitoring and diseases.

California Polytechnic State University Animal Science Department
http://www.calpoly.edu/~asci/
Contains links to areas of general interest.

Center for Animal Health and Productivity
http://cahpwww.nbc.upenn.edu/
Contains links and information relating to aquaculture, dairy cattle, nutrient management, poultry, swine, veterinary medicine, etc.

Contech - Intelligent Animal Control
http://www.scatmat.com
A source of information on high tech pet and animal training and repelling products for vets, behaviorists, trainers, gardeners, farmers and pet owners.

Council for Agricultural Science and Technology (CAST)
http://www.netins.net/showcase/cast/naftmenu.htm
This site contains a report summary of NAFTA, as well as news releases, a list of task force members, and a link to the complete NAFTA document.

Environmental Regulations for Livestock Facilities
http://gaia.ageng.umn.edu/extens/ennotes/enwin95/regs.html
This site contains U.S. EPA regulations for livestock producers.

Fact Sheet and Current Reports from Oklahoma State
http://www.ansi.okstate.edu/exten/publica.html
This site contains publications of general interest on topics such as agribusiness, beef cattle, dairy, general animal science, horse, poultry, sheep, and swine.

International Meat & Poultry HACCP Alliance Food Safety Information
http://ifse.tamu.edu/alliance/foodsafety.html

Kansas State University Animal Sciences and Industry Department
http://www.oznet.ksu.edu/dp_ansi/
This is an impressive page with links to animal breeding and swine science, among other useful links. It also contains links to the Department of Animal Sciences and Industry staff.

List of Livestock Breeds from Oklahoma State University
http://www.ansi.okstate.edu/breeds
This site contains information on cattle, donkeys, horses, goats, sheep, and swine, and is searchable by breed or region.

Livestock Information Sites
http://www.agric.gov.ab.ca/livestck/index.html
Provides a list of links to information on beef, dairy, poultry, and hogs, as well as general livestock information.

Livestock Publications from the Colorado Cooperative Extension Service
http://www.colostate.edu/Depts/CoopExt/PUBS/LIVESTK/publive.html
Contains documents on the following topics: feeding, handling and facilities, insects and diseases, livestock production, and poultry.

Livestock Section of WWW Virtual Library
http://www.ansi.okstate.edu/library/
Information concerning species, animal rights, markets, fairs and expos, and academic conferences can be located here.

Livestock World
http://www.pitchfork.com

Market Reports for Livestock from the University of Kentucky
gopher://shelley.ca.uky.edu:70/11/agmkts/market_wire

This site contains market information, domestic and international prices and commentaries, state by state price listings, and more.

The Meating Place
http://www.mtgplace.com

Missouri By-Product Feed Prices
gopher://etcs.ext.missouri.edu:70/00/agebb/ansci/dairy/bullet1.r

Contains price information on alfalfa pellets, brewers grain, corn gluten feed, cottonseed, distillers grain, hominy, malts sprouts, and more.

Missouri Extension Publications
http://etcs.ext.missouri.edu/publications/xplor/index.htm

This site's topics include aging, animal sciences, horticulture, crops, communications, business and industry, farm management, human development, food science and nutrition, and water quality.

NAGRP News
http://probe.nalusda.gov:8000/otherdocs/nagrpnews/nagrpnews11.html

Contains national animal genome research program news.

National Animal Poison Control Center Information Page
http://www.napcc.aspca.org

The ASPCA National Animal Poison Control Center, an operating division of the American Society for the Prevention of Cruelty to Animals (ASPCA), is the first national animal-oriented poison control center in the United States.

Pellcom Online Livestock Marketplace
http://www.pellcom.com

Pollution Potential of Livestock Manure
http://gaia.ageng.umn.edu/extens/ennotes/enwin95/manure.html

University of Illinois College of Veterinary Medicine
http://www.cvm.uiuc.edu
This page includes positions available in the department, announcements, reports, educational resources, reference desk for finding resources, and more.

University of Kentucky Animal Sciences Department
http://www.uky.edu/Agriculture/AnimalSciences/ukdas.html
Contains information on research programs in nutritional and ruminal microbiology, and equine nutrition and exercise physiology.

University of Missouri-Columbia Animal Sciences Department
http://www.asrc.agri.missouri.edu
Contains general departmental links and course information.

Virginia Tech Animal Sciences Department
gopher://gopher.ext.vt.edu:70/11/vce-data/aps
At this site you can search all animal and poultry sciences files or access information on 4-H poultry, beef, horses, sheep, small/specialty poultry, and swine.

BEEF CATTLE

A*L*O*T Angus Associations Page O'Links
http://www.erinet.com/carl/alotlink.html

Beef Breeds and Associations
http://ops.agsci.colostate.edu/~scomstoc/

Beef Cattle Resources
http://www.ansi.okstate.edu/library/cattbeef.html
This site contains a number of links to breeding information and associations, central bull test stations, publications, commercial pages, disorders and pests, nutrition and feeding, etc.

Beef Industry Information System
http://www.ncanet.org/

Beef Information from Virginia Tech
gopher://gopher.ext.vt.edu:70/11/vce-data/aps/beef
This site contains an extensive list of articles on the topics of nutrition, health, genetics, reproduction, marketing/economics, end product, and management.

Beef Today Magazine
http://www.beeftoday.com/

Breeding, Feeding, and Marketing Information from Missouri
http://www.ext.missouri.edu/publications/xplor/agguides/ansci/beef.htm
Contains a catalog and summary of articles relating to breeding, reproduction, calving, rationing, feed activities, grains, backgrounding, disease prevention, etc.

Cattle Breeds
http://www.ansi.okstate.edu/breeds/cattle/
Includes a discussion of cattle breeds with their respective cattle associations and mailing addresses.

Cow Culling Decisions
http://ag.arizona.edu/AREC/cull/culling.html
The interactive Decision Support System (DSS) provides recommended culling decisions.

Cow Sense Herd Management Software
http://www.midwestmicro.com

Cow Town America!
http://www.cowtown.org/index.html

Eastern Breeders, Inc.
http://casper.ilms.com/ebi/

Electronic Zoo/Net Vet - Cows
http://netvet.wustl.edu/cows.htm
This site has a list of links relating to beef, dairy, and commercial cows.

EPA/USDA Ruminant Livestock Methane Program
http://www.epa.gov/docs/GCDOAR/ruminant.html
Includes a discussion of the background of the Ruminant Livestock Methane Program, as well as how to get involved, and the benefits of involvement.

Fitzpatrick Cattle Company
http://rfitz.com/cattle

Limousin World Magazine
http://www.agrione.com/LimousinWorld

The National Cattlemen's Beef Association
http://www.beef.org

Precision Beef Alliance
http://www.public.iastate.edu/~magico/pba.html
Cattle producers based in southwest Iowa dedicated to providing information services to members as well as helping to forward the beef industry as a whole.

Silver Creek Feeders, Inc.
http://www.silvercreekfeeders.com

Texas Beef Council
http://www.txbeef.org

U.S. Belted Galloway Society
http://www.beltie.org

BEEF RELATED LISTSERVS

Agric-L
Subscribe to: listserv@uga.cc.uga.edu

Beef-L
Subscribe to: listproc@listproc.wsu.edu

BeefToday-L
Subscribe to: majordomo@angus.mystery.com

Graze-L
Subscribe to: listserv@taranaki.ac.nz

Forage-MG
Subscribe to: almanac@oes.orst.edu

DAIRY

Agriculture Marketing Service Dairy Division
http://www.usda.gov/ams/dairy.htm
Describes dairy programs, services, and resources. The site includes a list of available publications, key contacts, and employment opportunities.

California Milk Advisory Board
http://www.calif-dairy.com/

California Polytechnic State University Dairy Science Department
http://www.calpoly.edu/~dsci/

Center for Dairy Profitability
http://www.wisc.edu/dairy-profit

Cow Town America!
http://www.cowtown.org/index.html

Dairy Information from Cornell
http://www.cce.cornell.edu/programs/ag/dairy.html

Dairy Information from Missouri
http://www.ext.missouri.edu/publications/xplor/agguides/dairy/index.htm
Dairy breeding, feeding, management, and marketing information are available at this extension site.

Dairy Cattle Resources
http://www.ansi.okstate.edu/library/dairy/
Contains information on breeds, breed associations, and publications.

DHIA
http://www.dhia.org

Eastern Breeders, Inc.
http://casper.ilms.com/ebi/

Electronic Zoo/NetVet - Cows
http://netvet.wustl.edu/cows.htm

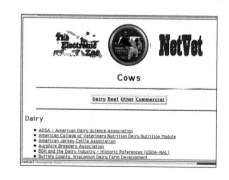

National Dairy Database
http://www.inform.umd.edu/EdRes/Topic/AgrEnv/ndd

Soyplus-Gold Standard
http://www.soyplus.com

GOATS

Alberta Goat Breeders Association
http://www.freenet.edmonton.ab.ca/agba/

Electronic Zoo/NetVet - Small Ruminants
http://netvet.wustl.edu/smrum.htm
Information on goats, goat resources, dairy goat production review, and raising goats is located here.

Goat Information from Irvine Mesa Charros 4-H Club
http://www.ics.uci.edu/~pazzani/4H/Goats.html
This page contains information relating to every aspect of goat development, as well as a large listing of other goat resources.

Goat Resources from Oklahoma State
http://www.ansi.okstate.edu/library/goats.html
Contains information on husbandry and production, diseases, parasites, and drugs.

GoatWeb!
http://www.goatweb.com

OSU Goat Information
http://www.ansi.okstate.edu/library/goats.html

USDA National Goat Handbook
http://www.inform.umd.edu:8080/EdRes/Topic/AgrEnv/ndd/goat/

HORSES

Arabian Horse World Magazine
http://www.ahwmagazine.com
Contains information on breeders, sales, studs, AHW information and show results.

The Bridal Path Network
http://www.bridlepath.net/
Lists horse related sites by state, breed, discipline and services.

Churchill Downs Home Page
http://www.kentuckyderby.com
This site covers Hugh Finn's horse racing news, the current meet, simulcast wagering, track history, the Kentucky Derby, and more.

Electronic Zoo/Net Vet - Horses
http://netvet.wustl.edu/horses.htm

Horse Farm Management
http://web.profiles.com

Horse Resources from Oklahoma State
http://www.ansi.okstate.edu/library/equine.html
Contains information regarding breeds, diseases, disorders, parasites, etc.

HorseNet
http://www.horsenet.com/
Contains information on vacation and travel, stores, associations and organizations, veterinary and health care, magazines, and the HorseNet Classifieds.

HorseWeb
http://www.horseweb.com/
This site contains Web pages of horses, horse products and services.

International Museum of the Horse in Lexington, KY
http://www.horseworld.com/imhmain.html
This site contains historical information on the horse, including exhibits from the International Museum of the Horse and contributing museums, information on the Kentucky Horse Park, and other recommended equestrian and museum resources.

On Eagle's Wings Equestrian Center
http://members.aol.com/mariehorse/index.htm

Show Horse Showplace
http://www.showhorse.com/
This site has horse show results, pictures and motion video of winners at recent horse shows, horses for sale, and other horse related items.

The (unofficial) American Saddlebred Horse
http://saddlebred.agriequine.com
This site contains information about the American Saddlebred horse breed, including the Saddlebred Art Gallery, and the official site of The American Saddle Horse Museum.

University of Missouri Xplor Publications
http://www.ext.missouri.edu/publications/xplor/agguides/ansci/horses.htm
Contains a catalog and summary of articles on horse related topics including breeding, genetic, feeding, and other information.

The Wild Horse, Mustang, and Burro Page
http://iquest.com/~jhines/mustang/
This site has information about wild horses, including how to adopt a wild horse or burro, mustang organizations, mustang news, and other wild horse sites.

Horse Related Listservs

Equine-L
Subscribe to: listserv@psuvm.psu.edu

Horse Related Newsgroups
news:rec.equestrian

OTHER LIVESTOCK

Bobkat Llamas
http://www.techline.com/~bobkat/

Castalia Llamas
http://www.rockisland.com/~castalia/cllama.html

Cotton-Packin' Llamas!
http://cessna.tippecanoe.com/cotton.htm
This site has information on llama training, breeding, care and marketing, sales list, packing information, general information, and pictures.

Crocker's Ostrich Page
http://www.islandnet.com/~ski/ostrich/canost.html
Links to breeders and people in the ostrich industry, research in the industry, the Oklahoma State Ostrich Book, Canadian Ostrich Magazine, etc.

The Double B Ranch
http://llamasales.com

Hillview Llamas
http://www.owt.com/hillview

The International Lama Registry
http://www.vtown.com/ilr
The ILR is the largest compilation of lama genealogical information in the world.

Kent Laboratories - Veterinary Diagnostic Division
http://www.rockisland.com/~newmoon/vet.html

Llama Information
http://www.webcom.com/~degraham/
Contains everything from games to veterinary information.

Llama Owners of Washington State
http://www.rockisland.com/~newmoon/lows.html

Llama Rose Farm
http://www.silverlink.net/llamarose

Mariko Llamas
http://people.infospeedway.net/~mnjdoyle

Ostrich-Emu Farming Software
http://www.imwa.com.au

The Ostrich-Emu InfoNet Home Page
http://www.ostrich-emu.com
This site contains contact information, as well as links to associations and organizations, events, equipment vendors, processors, co-ops, restaurants & delis, etc.

SauerMugg's Big Bird Ranch
http://www.pier37.com/sauer/
Includes links to emu origin and history, purchasing birds, diseases and illness, emu associations and organizations, publications, and insurance companies.

Virtual Emu by the Royal Australasian Ornithologists Union
http://www.vicnet.net.au/~raou/raou.html
Contains information about the red-tailed black-cockatoo appeal, Australian bird images, events, research projects, RAOU bird observatories, etc.

The Water Buffalo Homepage
http://ww2.netnitco.net/users/djligda/waterbuf.htm

POULTRY

Ag Agent Handbook - Poultry
http://hammock.ifas.ufl.edu/text/aa/39587.html
This site contains information about managing poultry, including articles on management of small flocks of chickens, diseases and parasites of poultry, structures for composting of broiler mortality, etc.

Agriculture Marketing Service, Poultry Division
http://www.ams.usda.gov/poultry/
This site contains information on poultry programs, services, and resources, publications, key contacts, and employment opportunities.

California Egg Commission
http://www.eggcom.com

California Poultry Industry Federation
http://www.ainet.com/cpif/

The Chicken Coop
http://www.transport.com/~lhadley/index.html
This site contains a directory of poultry enthusiasts on the WWW, as well as raising and breeding information, and other links.

The Chicken Page
http://ccwf.cc.utexas.edu/~ifza664/index.html
Contains chicken facts, diseases, nutrition, and breed information.

Electronic Zoo/NetVet - Birds
http://netvet.wustl.edu/birds.htm
Contains a list of links relating to poultry and pet birds.

NCSU Poultry Science Department's Home Page
http://www2.ncsu.edu/ncsu/cals/poultry/
This site contains information about the department, students, interesting places for agriculture, birds, biology, and more.

Texas A&M Poultry Science Department
http://gallus.tamu.edu

U.S. Poultry Gene Mapping
http://poultry.mph.msu.edu
Contains information about chicken biology and gene mapping.

SHEEP

American Texel Sheep Association
http://www.texel.org

The Bandana Sheep Company
http://www.bandanasheep.com

Electronic Zoo/NetVet - Small Ruminants
http://netvet.wustl.edu/smrum.htm
Contains information on breeds, diseases, and parasites.

Getting Started in Sheep
http://www.gov.sk.ca/agfood/farmfact/lis5447.htm

Sheep Homepage
http://members.aol.com/culhamb/sheepweb.htm

Sheep Information from Michigan State University
http://www.canr.msu.edu/dept/ans/sheep/sh2.htm
Contains information on the MSU sheep flock and sheep sale.

Sheep Information from Virginia Tech
gopher://gopher.ext.vt.edu:70/11/vce-data/aps/sheep
Contains articles on nutrition, health, genetics, reproduction, marketing/economics, end product, and management.

Sheep on the Web
http://members.aol.com/culhamb/sheepweb.htm

Sheep Resources from Oklahoma State
http://www.ansi.okstate.edu/library/sheep.html
Information on breeds, publications, diseases, disorders, and pests can be found here.

University of Missouri Xplor Publications - Sheep
http://www.ext.missouri.edu/publications/xplor/agguides/ansci/sheep.htm
Includes a catalog and summary of articles relating to sheep.

SWINE

Ag Agent Handbook
http://hammock.ifas.ufl.edu/text/aa/39589.html
This site contains information about planning swine facilities, selection and mating of breeding stock, feeding, and swine health program.

Dean Houghton's Unplugged in Polo
http://ourworld.compuserve.com/homepages/Dean_Houghton/
This site belongs to Dean Houghton, the editor of Hogs Today magazine.

Electronic Zoo/NetVet - Pigs
http://netvet.wustl.edu/pigs.htm
This site has a lengthy list of pig-related links, covering topics from images to breeding to disease.

Farm Journal's Hogs Today
http://www.hogstoday.com

National Pork Producers Council
http://www.nppc.org

North Carolina State University Swine Extension
http://jah.asci.ncsu.edu
Contains nutrition and feeding topics, economic and resource topics, waste management, water quality and odor topics, reproduction topics, and more.

Ohio Pork Producers Council
http://www.ohiopork.org

Pig Disease Information Center at the University of Cambridge
http://www-pdic.vet.cam.ac.uk/
Contains news pages, expert technical information about swine health and production, databases of swine information resources, a directory of swine consultants, and a swine discussion forum for veterinarians.

Prairie Swine Centre
http://www.lights.com/psc
The Centre is the largest swine production research facility in Canada, carrying out intensive research, education and technology transfer activities.

Swine Information from Missouri
http://www.ext.missouri.edu/publications/xplor/agguides/ansci/swine.htm
This site contains breeding, feeding, health, and management information. Scroll down until you see Animal Science-Swine information.

Swine Information Site from Virginia Tech
gopher://gopher.ext.vt.edu:70/11/vce-data/aps/swine
This site contains articles on swine nutrition, health, genetics, reproduction, marketing/economics, end product, and management.

Swine Resources from Oklahoma State
http://www.ansi.okstate.edu/library/swine.html
Information on breeds, diseases, parasites and nutrition are located here.

U.S. Pig Gene Mapping
http://www.public.iastate.edu/~pigmap/pigmap.html

MAILING LISTS ON THE INTERNET

AG-EXP-L - Ag Expert Systems
Subscribe: listserv@vm1.nodak.edu

Clark Consulting International, Inc.
http://www.interaccess.com/consulting/maillist.html
Contains up-to-date lists of listserv mailing list groups.

PamL Mailing List
http://www.neosoft.com/internet/paml

Ruraldev (Community and Rural Economic Development)
Subscribe: listserv@ksuvm.ksu.edu

Tile.Net/Listserv Page
http://www.tile.net/tile/listserv/index.html

Veterinary and Animal LISTSERV Archive
http://netvet.wustl.edu/vmla.htm
Includes listserv log files.

MANAGEMENT AND MARKETING

Agricultural Labor Management from UC Berkeley
http://www.cnr.Berkeley.edu/ucce50/7grisha.htm
Contains information, links and resources for agricultural labor management.

Clark Consulting International (International Agriculture Marketing Consultants)
http://www.interaccess.com/consulting/
Provides agricultural marketing consulting services, including marketing communications planning, market research, public relations, advertising, etc.

Commercial Farming Regulations
http://hammock.ifas.ufl.edu/txt/fairs/19553
Lists regulations for farm employers and employees.

Farm Business and Farm Management - FBFM
http://www.ag.uiuc.edu/~fbfm/fbfm.html
FBFM is a cooperative education program to assist farmers with decision making in business and family records.

Farm Management By Nerd World Media(TM)
http://www.nerdworld.com/nw678.html
Contains a searchable index of sites related to farm management.

Farm Management, Nebraska Extension Publications
http://ianrwww.unl.edu/ianr/pubs/catalog/farmgmt.htm
A catalog of publications and computer programs dedicated to farm management.

FarmMARKET
http://www.hort.purdue.edu/newcrop/farmmarket/farmmarket.html
This site lists farmer's markets in each state.

International Agricultural Trade Information Service
http://caticsuf.csufresno.edu:70/1/atinet/agtrade

The International Farm Management Association
http://www.re.ualberta.ca/ifma/begin.htm

MARKET AND PRICE INFORMATION SITES

Agricultural Market Advisory Services
http://www.aces.uiuc.edu/~agmas/

Agricultural Market Information Virtual Library
http://www.aec.msu.edu/agecon/fs2/market/contents.htm

Agriculture Marketing Service (USDA)
http://www.ams.usda.gov/
Contains news and information from the Agriculture Marketing Service.

AgriGator Ag Market News
http://www.ifas.ufl.edu/www/agator/htm/agmarket.htm

Asian Regional Agribusiness Project
http://www.milcom.com/rap
Provides market and technical information to Asian and U.S. agri-businesses.

Internet Addresses

Brown County Co-op
http://www.bcca.net

Chicago Board of Trade
http://www.cbot.com

Chicago Board Options Exchange
http://www.cboe.com

Chicago Mercantile Exchange
http://www.cme.com

CNN/fn - Commodities
http://cnnfn.com/markets/commodities.html

The Coffee, Sugar & Cocoa Exchange
http://www.csce.com

DTN Farmdayta Online
http://www.dayta.com

Farm Journal's Market Information
http://www.farmjournal.com

Farmer's Co-op Elevator
http://www.bbc.net/co-op

Futures and Options Trading Group
http://www.teleport.com/%7Efutures/

Futures Market Commentary
http://www.trade-futures.com/reports.html
Provides weekly futures trading information from MBH commodity advisors.

The Kansas City Board of Trade
http://www.kcbt.com/

Livestock Marketing Information Center
http://lmic1.co.nrcs.usda.gov
Provides economic projections to the livestock industry.

Livestock Reports from the University of Nebraska
http://ianrwww.unl.edu/MARKETS/LIVESTCK.HTM
This site contains various articles of interest on the following topics: beef, dairy, horses, pork, sheep, and general agriculture topics.

The Market Analysis Division Website
http://www.agr.ca/policy/winn/biweekly/English/index2e.htm
Provides market information, analysis and forecasting of supply, demand, trade and prices to industry and governments.

Market News via OSU
http://ag.arizona.edu/AREC/mnews/Amarket.html

Market Reports from Mississippi State
http://www.ces.msstate.edu/market/
Cheese, vegetables, nuts, eggs, and cream market information resides on this server. Includes information for Mississippi, its neighbors, and the nation.

The Mid-America Commodity Exchange
http://www.midam.com/

Minneapolis Grain Exchange
http://wwwmgex.com

Missouri By-Product Feed Prices
http://etcs.ext.missouri.edu:70/0/agebb/ansci/dairy/bullet1.r
Hominy, cotton seed hulls, corn gluten feed, and alfalfa pellets are just a few of the by-products listed on this page.

NASS Reports from USDA
http://ag.arizona.edu/AREC/mnews/NASS.html

New York Cotton Exchange
http://www.nyce.com

New York Mercantile Exchange
http://www.nymex.com/

Prices Received by Commodity & Historic Data Series
http://www.mannlib.cornell.edu/cgi-bin/description.cgi?529.html

Successful Farming's Agriculture Online Markets
http://www.agriculture.com/markets/mktindex.html
Successful Farming's markets page has links to articles on marketing strategy, reports, and market analysis, as well as links to current market prices.

TFC Commodity Futures & Financial Market Charts
http://www.tfc-charts.w2d.com/

Today's Market Prices
http://www.todaymarket.com/
Provides wholesale fruit and vegetable market prices from the United States, Canada, Mexico, Europe, Asia and Latin America, classified by product, terminal, varieties and sizes.

USDA's Daily Cash Grain Reports from the University of Kentucky
gopher://shelley.ca.uky.edu:70/11/agmkts/market_wire/grain
Includes market reports for many states, including Kentucky.

USDA's Daily Future Grain Reports from the University of Kentucky
gopher://shelley.ca.uky.edu:70/00/.agwx/usr/markets/usda/MSGR711

USDA's Market Wire News Reports from the University of Kentucky
gopher://shelley.ca.uky.edu/11/agmkts/market_wire
Contains market reports on dairy, feedstuffs, fruits and vegetables, grain, hay, livestock, meat, poultry, and tobacco. Also has futures and international reports.

U.S. Economy: Business Cycle Indicators
http://www.cris.com/~netlink/bci/bci.html

Useful Technologies
http://www.usefultech.com
Contains information on farm management and hedging and bringing modern financial markets hedging skills to the agricultural community.

PESTICIDES

Environmental Working Group
http://marlon.ewg.org

EPA Pesticide Regulation Notices
http://www.epa.gov/opppmsd1/PR_Notices

National Pesticides Telecommunications Network
http://ace.orst.edu/info/nptn/factshts.htm

Pesticide Handling and Storage Tutorial
http://pasture.ecn.purdue.edu/~embleton/pest/start.html

Pesticide Operational Area Containment Rules
http://www.state.sd.us/state/executive/doa/das/pest_oac.htm

Pesticide Record Keeping Requirements
http://hammock.ifas.ufl.edu/txt/fairs/42982

Pesticide Storage and Handling
http://waterhome.tamu.edu/texasyst/farmworkbooks/faswork2.html

PRECISION FARMING

AgLeader
http://www.agleader.com

Agri Alternatives
http://www.agrialt.com

@griculture Online
http://www.agriculture.com/technology/index.html

Agri-Growth
http://www.agrigrowth.com

Agri-Logic
http://www.agrilogic.com

Agris
http://www.agris.com

Ashtech
http://www.ashtech.com

Association of Ag Computing Companies
http://asae.org/aacc

Cartography Resources: Commercial
http://geog.gmu.edu/gess/jwc/cartogrefs.html

Case-IH/Advanced Farming Systems
http://www.casecorp.com/agricultural/afs/index.html

Centre for Precision Farming
http://www.silsoe.cranfield.ac.uk/cpf

Communication Systems International
http://www.csi-dgps.com

Differential Corrections, Inc.
http://www.dgps.com

Emerge Field Imaging
http://www.emerge.wsicorp.com

ESRI (GIS)
http://www.esri.com

Farmers Software Association
http://www.farmsoft.com

GeoSystems
http://www.geosys.com

GIS Data for Northern California
http://www.pacific.net/~cbrooks/gis1.htm

GIS WWW Resource List
http://www.geo.ed.ac.uk/home/giswww.html

Global Positioning System Overview
http://www.utexas.edu/depts/grg/gcraft/notes/gps/gps.html

The GPS Home Page
http://www.technologyplus.com/gps

Great GIS Information Sites!
http://www.hdm.com/gis3.htm

Growmark
http://www.growmark.com

Hammond Map Co.
http://www.hammondmap.com/

i3 - Information Integration & Imaging
http://www.i3.com/

ImageNet
http://www.coresw.com/

Introduction to GPS Applications
http://galaxy.einet.net/editors/john-beadles/introgps.htm

Itronix
http://www.itronix.com

John Deere Precision Farming
http://www.deere.com/greenstar

The Living Earth, Inc.
http://livingearth.com/

Magellan Geographix - Digital Maps and Cartography
http://www.magellangeo.com/

MapQuest!
http://www.mapquest.com/

Omnistar
http://www.omnistar.com

Partners in Precision
http://www.smartfarm.com

Precision Farming Institute
http://pasture.ecn.purdue.edu/~mmorgan/PFI/graphic.html

Precision Farming Megalinks
http://nespal.cpes.peachnet.edu/pf

Precisionag.com
http://www.precisionag.com

Rockwell/VISION
http://www.cacd.rockwell.com/bus_area/ag_sys/index

Satloc Precision GPS Applications
http://www.klasi.com/index.html

Spectrum Technologies
http://www.specmeters.com

SST Development Group
http://www.sstdevgroup.com

Starlink, Inc.
http://www.starlinkdgps.com

Successful Farming Magazine/Precision Ag Messages
http://www.agriculture.com/agtalk/Precision_Agriculture/listmsgs.cgi

Terra Industries/Precision Ag
http://www.terraindustries.com/precision/index.html

Tyler
http://www.teamtyler.com

U.S. Coast Guard Navigation Center
http://www.navcen.uscg.mil

SOIL AND WATER

Colorado Water Resources Research Institute
http://www.ColoState.EDU:80/Depts/CWRRI/
Contains a link to the Colorado Water Newsletter, as well as links to CWRRI's research program and the National Institute for Water Resources.

Hydroponics
http://www.growroom.com/HydroList/index.html

Indiana Water Resources Research Center
http://ce.ecn.purdue.edu/wrrc.html
Contains information about the Center, the National Institute for Water Resources, Indiana WETnet, IWRRC current projects, and IWRRC publications.

Inter-Urban Water Farms Online
http://www.ilri.nl/lswlinks.html

Land, Soil and Water Internet Resources
http://www.ilri.nl/lswlinks.html
A selection of key sources on the Internet of interest to researchers in the field of land, soil and water.

Oregon State University Crop and Soil Science Department
http://www.css.orst.edu/
Links to the Forage Information System, on-line classes, calendar of events, research, and information about the department reside on this homepage.

Purdue University Agricultural Extension Service
http://hermes.ecn.purdue.edu:8001/http_dir/acad/agr/extn/extn.html
Includes the National Water Quality Database.

TWRI Texas Waternet
http://twri.tamu.edu
Contains information and links about Texas water resources and research.

University of Delaware College of Agricultural Sciences
http://bluehen.ags.udel.edu/
Contains many links, including a search of their AGINFO plant database and Delaware Water Resources Information System.

University of Minnesota Soil, Water, and Climate Department
http://www.soils.agri.umn.edu/
This site has general departmental information, as well as a compilation of links to agricultural information and general information of use on the Web.

University of Missouri Soil and Water Information
http://www.ext.missouri.edu/publications/xplor/agguides/soils/index.htm

USDA's Water Management Research Laboratory
http://asset.arsusda.gov/wmrl/wmrl.html

Water Resources Center at the University of Wisconsin-Madison
http://www.library.wisc.edu/libraries/Water_Resources/page.htm
This site will let you access the Water Resources Center Library, now online via telnet, and the Water Resources Center staff directory.

STATE DEPARTMENTS OF AGRICULTURE

Although states not listed did not have a home page as of the printing date, all were in the process of building a home page or had plans to create a home page in the near future. Check our Web site for updates as they become available.

Alaska Department of Natural Resources—The Division of Agriculture
http://www.dnr.state.ak.us/ag/index.htm

Arizona Department of Agriculture
http://agriculture.state.az.us

California Department of Food and Agriculture
http://www.cdfa.ca.gov

Colorado Department of Agriculture
http://www.ag.state.co.us

Connecticut State Department of Agriculture
http://www.state.ct.us./doag

Florida Department of Agriculture and Consumer Services Homepage
http://www.fl-ag.com

Georgia Department of Agriculture Homepage
http://www.agr.state.ga.us

Hawaii Department of Agriculture Homepage
http://www.mic.hawaii.edu/hawaiiag/212.htm

Idaho Department of Agriculture
http://www.agri.state.id.us

Illinois Department of Agriculture
http://www.state.il.us/agr

Indiana Office of The Commissioner of Agriculture
http://www.ai.org/oca

Iowa Department of Agriculture Homepage
http://www.state.ia.us/agriculture/index.html

Kansas Department of Agriculture
http://www.ink.org/public/kda

Kentucky Department of Agriculture
http://www.state.ky.us/agencies/agr/kyagr.htm

Louisiana Department of Agriculture
http://www.ldaf.state.la.us/

Maine Department of Agriculture, Food and Rural Resources Homepage
http://www.state.me.us/agriculture/homepage.htm

Maryland Department of Agriculture
http://www.mda.state.md.us

Massachusetts Department of Food and Agriculture
http://www.massgrown.org

Michigan Department of Agriculture
http://www.mda.state.mi.us/

Minnesota Department of Agriculture
http://www.mda.state.mn.us

Mississippi Department of Agriculture
http://www.mdac.state.ms.us/

Montana Department of Agriculture
http://161.7.66.167/index.htm

National Association of State Departments of Agriculture - NASDA
http://www.nasda-hq.org/nasda/nasda/index1.htm
Supports and promotes the American agriculture industry, while protecting consumers and the environment through the development, implementation, and communication of sound public policy and programs.

Nebraska Department of Agriculture
http://www.agr.state.ne.us

Nevada Department of Agriculture
http://www.state.nv.us/busi_industry/ad/index.htm

New Hampshire Department of Agriculture, Markets and Food
http://www.state.nh.us/agric/aghome.html

New Jersey Department of Agriculture
http://www.state.nj.us/agriculture

▼ *Part II*

New Mexico Department of Agriculture Homepage
http://nmdaweb.nmsu.edu

New York State Department of Agriculture and Markets
http://unix2.nysed.gov/ils/executive/agric/agric.htm

North Carolina Department of Agriculture & Consumer Services
http://www.agr.state.nc.us

North Dakota Department of Agriculture
http://www.state.nd.us/agr

Ohio Department of Agriculture
http://www.state.oh.us/agr

Oklahoma Department of Agriculture Homepage
http://www.state.ok.us/~okag/aghome.html

Oregon Department of Agriculture
http://www.oda.state.or.us/oda.html

Pennsylvania Department of Agriculture
http://www.pda.state.pa.us

South Carolina Department of Agriculture
http://www.state.sc.us/scda

South Dakota Department of Agriculture
http://www.state.sd.us/state/executive/doa/doa.html

Tennessee Department of Agriculture
http://www.state.tn.us/agriculture

Texas Department of Agriculture
http://www.agr.state.tx.us

Utah Department of Agriculture
http://www.ag.state.ut.us

Vermont Department of Agriculture, Food and Markets
http://www.cit.state.vt.us/agric/index.htm

Virginia Department of Agriculture and Consumer Services Homepage
http://www.state.va.us/~vdacs/vdacs.htm

Wyoming Department of Agriculture Homepage
http://wyagric.state.wy.us/

TURF MANAGEMENT

CUE Turf Management
http://www.nps.gov/cue.html

NCSU Plant Disease Information Notes
http://www.ces.ncsu.edu/depts/pp/notes/Turfgrass/turfgrass_contents.html
Contains information about turf diseases and problems.

Ornamental and Turf Audiovisual
http://ipcm.wisc.edu/PAT/Categories/Ornamental/audiovisual.htm
Produced in April 1995, the Ornamental and Turf video contains the presentations given at a commercial applicator training session. The video is for sale.

Professional Lawn Care Association of America: Research
http://www.plcaa.org/abstract.html

Rutgers Professional Golf Turf Management School
http://www.rci.rutgers.edu/~ajenkins/turfbro.htm
This site is all about the program and includes an online admission application.

Sports Turf Association
http://www.uoguelph.ca/GTI/guest/sta.htm

Turf Grass Science - U of I
http://www.turf.uiuc.edu/

Turf Information from Cornell
http://www.fm.cornell.edu/grounds/turf.html
Contains many links to turf related topics, research projects and educational programs.

Turf Job Mart
http://www.uoguelph.ca/GTI/bb/tmbb3.htm
This is a spot for posting turf management jobs available/jobs wanted.

Turf Management
http://ag.arizona.edu/Ext/special_projects/plants/turf.html
Provides information about the Turf Management program at the University of Arizona.

Turf Management Construction
http://www.uoguelph.ca/GTI/bb/tmbb5.htm
This is a spot for posting questions, answers, etc., about turfgrass construction (golf, sports turf, landscape).

Turf Management Cultivation
http://www.uoguelph.ca/GTI/bb/tmbb11.htm
This is a spot for posting questions, answers, etc., about turfgrass cultivation: aerification, coring, and verticutting.

Turf Management Diseases
http://www.uoguelph.ca/GTI/bb/tmbb15.htm
This is a spot for questions, answers, etc., about turfgrass diseases and disease management.

Turf Management Establishment/Renovation
http://www.uoguelph.ca/GTI/bb/tmbb6.htm
This is a spot for posting questions, answers, etc., about turfgrass establishment or renovation: seeding, overseeding, sodding.

Turf Management Fertility
http://www.uoguelph.ca/GTI/bb/tmbb8.htm
This is a spot for posting questions, answers, etc., about turfgrass fertility.

Turf Management Irrigation
http://www.uoguelph.ca/GTI/bb/tmbb9.htm
This is a spot for posting questions, answers, etc., about turfgrass irrigation.

USGA Green Section
http://www.usga.org/green/index.html
The United States Golf Association's (USGA) Green Section, founded more than seven decades ago, remains the nation's chief authority regarding impartial, authoritative information for turfgrass management. This site contains program information and photographs.

WILDLIFE

A Sportsmans Dream - Wildlife Photography
http://www.onlinecol.com/sd/indexe.htm

Alaska Fish and Wildlife Unit
http://zorba.uafadm.alaska.edu/iab/unit_index.html
The Alaska Cooperative Fish and Wildlife Research Unit is part of a nation-wide cooperative program, initiated in 1935, to promote research and graduate student training in the ecology and management of fish, wildlife and their habitats.

National Wetlands Inventory
http://www.nwi.fws.gov
Contains the U.S. Fish and Wildlife Service National Wetlands Inventory.

University of Maine Wildlife Ecology Department
http://wlm13.umenfa.maine.edu/w4v1.html
Contains information about ongoing projects, faculty research interests, available positions, directions to the campus, and other WWW sites of interest.

U.S. Fish and Wildlife Service
http://www.fws.gov/

Watchable Wildlife in the National Parks
http://www.aqd.nps.gov/natnet/wv/watchwl.htm

ARTS AND SCIENCE

Andy Warhol Museum
http://www.clpgh.org/warhol

Archiving Early America
http://www.earlyamerica.com
This site contains a unique collection of documents from 18th Century America.

ArtsUSA
http://www.artsusa.org
This is the WWW site of the American Council for the Arts.

Cincinnati Contemporary Arts Center
http://www.spiral.org

Complete Works of William Shakespeare
http://the-tech.mit.edu/Shakespeare/works.html

Eureka! The Museum for Children
http://ourworld.compuserve.com/homepages/Eureka_Museum/

The Franklin Institute Science Museum
http://sln.fi.edu/

Guggenheim Museum
http://www.guggenheim.org

The Illinois State Museum
http://www.museum.state.il.us/

Interactive Things To Do Hotlist from the Franklin Institute Science Museum
http://sln.fi.edu/tfi/hotlists/interactive.html

Le Louvre
http://www.louvre.fr

Liftoff
http://liftoff.msfc.nasa.gov/
An interesting site offered by NASA about aerospace science.

Literature Hotlist from the Franklin Institute Science Museum
http://sln.fi.edu/tfi/hotlists/literature.html

National Aeronautics & Space Administration - NASA
http://www.nasa.gov/

Physical Science Hotlist from the Franklin Institute Science Museum
http://sln.fi.edu/tfi/hotlists/space.html

Reel.com
http://www.reel.com

Science Museums Hotlist from the Franklin Institute Science Museum
http://sln.fi.edu/tfi/hotlists/museums.html

The Smithsonian
http://www.si.edu

Space Science Museums Hotlist from the Franklin Institute Science Museum
http://sln.fi.edu/tfi/hotlists/space.html

Ticketmaster
http://www.ticketmaster.com/

USC Interactive Art Museum
http://digimuse.usc.edu/museum.html

The World Wide Web Virtual Library of Museums
http://www.comlab.ox.ac.uk/archive/other/museums.html

BUSINESS AND FINANCE

Accutrade
http://www.accutrade.com
Online trading.

Amazon.com Bookstore
http://www.amazon.com

American Stock Exchange
http://www.amex.com/

AMEX Financial Direct
http://www.americanexpress.com/direct
Online trading.

Arizona Stock Exchange
http://www.azx.com

AT&T 800 Directory
http://www.tollfree.att.net/dir800

Better Business Bureau
http://www.igc.apc.org/bbb

Brainwave
http://www.n2kbrainwave.com

Business and Economy List from Yahoo
http://www.yahoo.com/Business_and_Economy/

Capital Data
http://www.capitaldata.com

Capital NET
http://www.capn.com

Chicago Board of Trade
http://www.cbot.com

Chicago Board Options Exchange
http://www.cboe.com

Chicago Mercantile Exchange
http://www.cme.com

Chicago Stock Exchange
http://www.chicagostockex.com

CNNfn
http://www.cnnfn.com

Coffee, Sugar & Cocoa Exchange, Inc
http://www.csce.com

Datek Online
http://www.datek.com
Online trading.

Dow Jones Business Information Services
http://bis.dowjones.com/

Emerging Markets Companion
http://www.emgmkts.com

E-Schwab
http://www.eschwab.com
Online trading.

E*Trade
http://www.e-trade.com
Online trading.

Fannie Mae
http://www.fanniemae.com

FedEX Tracking
http://www.fedex.com/track_it.html

Fidelity WebXpress
http://personal.fidelity.com/trade
Online trading.

Financial Services Technology Consortium
http://www.fstc.org/

Financial Times
http://www.ft.com

Fortune Online Magazine
http://www.pathfinder.com/@@Vs5u*wcAjQPBJsnq/fortune/

Futures & Options Trading Group
http://www.teleport.com/~futures/

Futures Magazine Online
http://www.futuresmag.com

Global Charts
http://www.globalcharts.com/global.html

Global Financial Data
http://www.globalfindata.com

Global Investor Directory
http://www.global-investor.com

Inc. 500
http://www.inc.com/500
Inc. magazine's annual list of America's fastest-growing private companies.

International Finance Corporation
http://www.ifc.org

Investors Edge
http://www.irnet.com

InvestorWEB
http://www.investorweb.com
Investment information on companies and other news.

Job Trak
http://www.jobtrak.com/profiles

Kansas City Board of Trade
http://www.kcbt.com

The Kansas City Federal Reserve Bank
http://www.kc.frb.org

London International Financial Futures and Options Exchange - LIFFE
http://www.liffe.com

Mid-America Exchange
http://www.midam.com/index.html

Minneapolis Grain Exchange
http://www.mgex.com

Money Online
http://www.pathfinder.com/@@Vs5u*wcAjQPBJsnq/money/

Money Online Quick Quotes
http://quote.pathfinder.com/money/quote/qc
Here, you can get a stock quote online.

Money Talks
http://www.talks.com

Moody's Investors Service
http://www.moodys.com

Mutual Fund Home Page
http://www.brill.com/fundlink.html

Mutual Fund Lists from Yahoo
http://www.yahoo.com/Business_and_Economy/Companies/Financial_Services/Investment_Services/Mutual_Funds/

NASD Regulation, Inc.
http://www.nasdr.com

The Nasdaq Web Site
http://www.nasdaq.com/welcome.htm

National Association of Securities Dealers
http://www.nasd.com

New York Cotton Exchange
http://www.nyce.com

New York Mercantile Exchange
http://www.nymex.com

New York Stock Exchange
http://www.nyse.com

Numa Financial Systems
http://www.numa.com

PC Financial Network
http://www.pcfn.com

Philadelphia Stock Exchange
http://www.libertynet.org

Quick & Reilly
http://www.quick-reilly.com

Quick Quotes
http://quote.pathfinder.com/money/quote/qc

Quicken Homepage
http://www.quicken.com

Quote.com
http://quote.com

Resources for Economists on the Internet
http://econwpa.wustl.edu/other_www/EconFAQ/EconFAQ.html

San Diego Stock Exchange
http://www.sddt.com/data/data97/sdse/

Security First Network Bank, FSB
http://www.sfnb.com/

Standard & Poor's Rating Services
http://www.ratings.standardpoor.com

StockScreener
http://www.stockscreener.com
Screen stocks, sort results, research companies, check quotes and chart stocks.

Streetnet
http://www.streetnet.com
Online Investor's Guide.

Tax Forms and Instructions
http://www.fedworld.gov/ftp.htm

UPS Online
http://www.ups.com

U.S. Securities and Exchange Commission
http://www.sec.gov

Wall Street Net
http://www.netresource.com/wsn

The World Bank Homepage
http://www.worldbank.org

Yahoo! Finance
http://quote.yahoo.com

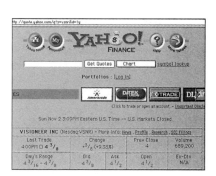

COMPUTER RESOURCES
COMPUTER MAGAZINES

Boardwatch Magazine Online
http://www.boardwatch.com

Byte
http://www.byte.com

C/Net Online
http://www.cnet.com

Cool Tool of the Day
http://www.cooltool.com

Computer Currents Interactive
http://www.currents.net

Ethernet Home Page
http://wwwhost.ots.utexas.edu/ethernet/ethernet-home.html

FutureNet
http://www.futurenet.co.uk

Internet.com
http://www.internet.com

MacAddict
http://www.macaddict.com

MacHome
http://www.machome.com

MacTech
http://www.mactech.com

MacToday
http://www.mactoday.com

MacUser
http://www.macuser.com
MacUser has merged with Macworld.

MacWEEK
http://www.macweek.com

MacWorld
http://www.macworld.com

The National Information Infrastructure Page
http://sunsite.unc.edu/nii/NII-Table-of-Contents.html

Netsurfer Digest
http://www.netsurf.com/nsd/index.html

PC Computing
http://www.pccomputing.com

PC Gamer
http://www.pcgamer.com

PC Magazine
http://www.zdnet.com/~pcmag/

PC Week
http://www.zdnet.com/~pcweek/

PCWorld
http://www.pcworld.com

PriceScan
http://www.pricescan.com
This site is a computer price search engine that helps take the hassle out of finding the best price on thousands of computer hardware and software products

WWWiz Magazine
http://wwwiz.com/

COMPUTER RETAILERS

Best Buy
http://www.bestnuy.com

Big Mac
http://www.wcn.com

Bottom Line
http://www.blol.com

CDW
http://www.cdw.com

Circuit City
http://www.circuitcity.com

Club Mac
http://www.club-mac.com

CompUSA
http://www.compusa.com

Computer Shopper
http://www.computershopper.com

ComputerCity
http://www.computercity.com

Cyberian Outpost
http://www.cybout.com

ICN
http://www.icni.com

LA Computer
http://www.lacc.com

Mac Works
http://www.macworks.com

Mac Zone
http://www.maczone.com

MacMall
http://www.macmall.com

MacWarehouse
http://www.warehouse.com/macwarehouse

Microwarehouse
http://www.warehouse.com

OnSale
http://www.onsale.com
Online auction.

PC Mall
http://www.pcmall.com

COMPUTER SOFTWARE

Adobe (win, mac)
http://www.adobe.com
Pagemaker, Photoshop, Sitemill, etc.

Autodesk (win, mac)
http://www.autodesk.com
Autocad.

Berkeley Systems (win, mac)
http://www.berksys.com
AfterDark, You Don't Know Jack and Triazzle.

Broderbund (win, mac)
http://www.broderbund.com
PrintShop.

Claris (win, mac)
http://www.claris.com
ClarisWorks, ClarisOffice, Filemaker Pro, Apple System Software, etc.

Connectix (win, mac)
http://www.connectix.com
Virtual PC, Ram Doubler and Speed Doubler.

Corel (win, mac)
http://www.corel.com
CorelSuite and Wordperfect.

Farallon (win, mac)
http://www.farallon.com
Timbuktu.

Forte Software
http://www.forteinc.com
Free Agent newsgroup reader.

Fractal Design (win, mac)
http://www.fractal.com
Painter and Ray Dream Studio.

Intuit (win, mac)
http://www.intuit.com
Quicken and QuickBooks.

Kodak
http://www.kodak.com
Includes a gallery of stunning 24 bit sample images.

Lotus
http://www.lotus.com

MacAMP (mac)
http://macamp.lh.net
MPEG3 player.

The Macintosh HyperArchive
http://hyperarchive.lcs.mit.edu/HyperArchive.html
This software archive contains links to nearly every Macintosh shareware application ever made.

Macromedia (win, mac)
http://www.macromedia.com
Freehand, xRes and Extreme 3D.

MacSoft (mac)
http://www.wizworks.com/macsoft

MetaCreations (win, mac)
http://www.metacreations.com
Fractal Design and MetaTools.

MetaTools (win, mac)
http://www.metatools.com

MetroWerks (win, mac)
http://www.metrowerks.com
Software development tools.

Microsoft (win, mac)
http://www.microsoft.com/products/default.asp
Whether it's utilities for Office '97 you want, or just a cool demo version of Fury3, Microsoft's got it.

Netscape Homepage (win, mac)
http://www.netscape.com
Download the newest versions of Netscape here.

Oklahoma State University Agronomy Department
http://clay.agr.okstate.edu/
Contains departmental information and information about agronomy software.

PriceScan
http://www.pricescan.com

Qualcomm (win, mac)
http://www.qualcomm.com
Eudora E-mail.

Quark (win, mac)
http://www.quark.com

Quarterdeck (win, mac)
http://www.quarterdeck.com
QEMM, CleanSweep and WebSTAR.

Shareware.com
http://www.shareware.com/
Shareware.com has one of the largest directories of downloadable shareware and freeware on the Web.

Symantec (win, mac)
http://www.symantec.com
Norton Utilities.

Texas A&M Extension Service Software Catalog
http://leviathan.tamu.edu
This site has downloadable software and information including over 2,300 clipart images, an experimental slides collection from the TAMU Department of Ag Communications, and much more.

WinAMP (win)
http://winamp.lh.net
MPEG3 player.

Windows95.com (win)
http://www.windows95.com
This is the best place get information about Windows 95 utilities and shareware.

WinZip (win)
http://www.winzip.com
Download an evaluation copy of the winzip shareware here.

E-ZINES - ELECTRONIC MAGAZINES

Apple Info Alley
http://survey.info.apple.com/ianetfinder/NetFinder.acgi

Apple Jedi
http://www.saracen.com/applejedi.html

Apple Wizards
http://www.erie.net/~iacas/applewizards

ATPM
http://www.atpm.com

Mac Daily Journal ($)
http://www.gcsf.com

Mac Fever
http://www.macfever.com

MacCom
http://www.maccom.net

MacReview
http://www.macreview.com

MacsRule
http://www.macsrule.com

My Mac
http://www.mymac.com

MyDesktop
http://www.mydesktop.com

TidBits
http://www.tidbits.com

WINSTUFF
http://www.winstuff.de

Win95 Annoyances
http://www.creativelement.com/win95ann
Excellent help site for intermediate Windows users to learn advanced topics. Contains solutions to various annoyances.

Win95 Tip Center
http://www.cam.org/~leonara/

HARDWARE COMPANIES

Adaptec
http://www.adaptec.com
SCSI cards.

Apple
http://www.apple.com

Asante
http://www.asante.com

ATI
http://www.atitech.com
Graphics cards.

Connectix
http://www.connectix.com
QuickCam.

CTX
http://www.ctxintl.com
Monitors.

Dell
http://www.dell.com
Computer systems.

Diamond Multimedia
http://www.diamondmm.com
Graphics cards, modems.

Farallon
http://www.farallon.com
Network cards and modems.

Gateway 2000
http://www.gw2k.com
Computer systems.

Global Village
http://www.globalvillage.com

Hewlett-Packard
http://www.hp.com
Printers.

IBM
http://www.ibm.com
PowerPC processor.

Iomega
http://www.iomega.com
ZIP & JAZ drives.

Matrox
http://www.matrox.com
Graphics cards, network hardware.

Motorola
http://www.mot.com
MacOS clones, modems, PowerPC processor.

Newer Technology
http://www.newertech.com
Processor upgrades, cache, RAM, etc.

PowerComputing
http://www.powercc.com
MacOS clones.

PowerTools
http://www.pwrtools.com
MacOS clones.

Sony
http://www.ita.sel.sony.com
Monitors, computers.

Toshiba
http://www.toshiba.com/tais/csd/products
Laptop and desktop computers.

UMAX
http://www.umax.com
Scanners, cameras, etc.

US Robotics
http://www.usr.com
Modems.

Wacom
http://www.wacom.com
Graphics tablets.

IBM COMPATIBLE COMPUTER RESOURCES

Barista Bill's IE4 and Windows Page
http://www.cleaf.com/~barista/

Windows 97 Information Site
http://www.win97.net/index.htm

Windows 98
http://www.activeie.com

INTERNET SERVICE PROVIDERS

Fiberlink Communications Corp.
http://www.fiberlinkcc.com
Telephone 1-714-788-2904.

jjj.net, Inc
http://www.jjj.net
Telephone 1-914-632-2271.

Netlimited
http://www.netlimited.net
Telephone 1-888-NET-LTD1.

UUNET Technologies
http://www.uu.net
Telephone 1-800-4UUNET4.

VPM Enterprises
http://www.vpm.com
Telephone 1-800-321-0221.

MACINTOSH RESOURCES

AbsoluteMac
http://www.absolutemac.com

C-Net Online
http://www.cnet.com

Digital Apple
http://www.digitalapple.com

The Fly On The Mac
http://www.mrmark.com/fly

INFO-MAC HyperArchive
http://hyperarchive.lcs.mit.edu/HyperArchive.html

InfoWorld
http://www.infoworld.com

Mac Resource Page
http://www.macresource.com

MacInsider
http://www.macinsider.com

The Macintosh Guide Book
http://www.everymac.com

MacinTouch
http://www.macintouch.com

MacOS Rumors
http://www.macosrumors.com

PowerBook Resource
http://ogrady.com/default.stm

PowerMac Resource Page
http://www.powermacintosh.com

SiteLink
http://www.sitelink.net
Best of the Web for Macintosh.

Version Tracker
http://www.versiontracker.com

Yahoo Mac Links
http://www.yahoo.com/Computers_and_Internet/Hardware/Personal_Computers/Macintosh

ECONOMIC DEVELOPMENT

Alaska's DCRA Economic Development Resource Guide
http://www.comregaf.state.ak.us/edrg_int.htm

Alberta Agriculture, Food and Rural Development (Ropin' the Web)
http://www.agric.gov.ab.ca
Provides access to information about most of Alberta's agriculture and food industry. Information about barley and other cereal crops as well as beef and forages is included.

The Appalachian Center at the University of Kentucky
http://www.uky.edu/RGS/AppalCenter

The Aspen Institute Rural Economic Policy Program
http://www.aspeninst.org/rural

Bureau of Labor and Statistics
http://stats.bls.gov/

Canadian Rural Information Service
http://www.agr.ca/policy/cris

Canadian Rural Restructuring Foundation
http://artsci-ccwin.concordia.ca/SocAnth/CRRF/crrf_hm.html

Census Bureau
http://www.census.gov

The Center for Civic Networking
http://www.civic.net:2401/ccn.html

Center for Economic Studies' Discussion Papers (Census Bureau)
http://www.census.gov/ces/papers.html

Center for Rural Massachusetts
http://www-unix.oit.umass.edu/~ruralma/CRM.html
Contains links to CRM publications and the Massachusetts Rural Development Council.

Centre for Rural Social Research
http://www.csu.edu.au/research/crsr/centre.htm

Economic Development Resources
http://www.state.ky.us/edc/otheredc.htm

Electric Power Research Institute
http://www.epri.com

Empowerment Zones and Enterprise Community Program
http://www.ezec.gov

Federal Domestic Assistance Catalog
http://www.gsa.gov/fdac

Forum for Applied Research and Public Policy
http://www.ra.utk.edu/eerc/forum/

Foundation for Rural Service
http://www.frs.org/

H-Rural WWW Site
http://h-net.msu.edu/~rural

The Foundation for Rural Service was established by the National Telephone Cooperative Association to inform and educate the public on the rural telecommunications industry and to improve the quality of life throughout rural America.

The HUD USER Homepage
http://www.huduser.org/
This is an information source for housing and community development researchers and policymakers.

Kentucky Assistance Online
http://www.rural.org/grants/

Mountain Association for Community Economic Development - MACED
http://www.maced.org

National Bureau for Economic Research
http://www.nber.com

National Resources Inventory Data
http://www.ncg.nrcs.usda.gov/nri.html

National Rural Development Partnership
http://www.rurdev.usda.gov/nrdp/index.html

Nebraska Department of Economic Development Homepage
http://www.ded.state.ne.us/index.html

Office of Social and Economic Data Analysis - OSEDA
http://www.oseda.missouri.edu

Ohio Business Retention and Expansion Program
http://www.ag.ohio-state.edu/~rande/

The Pratt Institute Center for Community and Environmental Development
http://www.picced.org

The Rural Advancement Foundation International
http://www.rafi.ca

The Rural Community College Initiative
http://www.mdcinc.org/rcci.html

Rural Development and Finance Corporation
http://www.rdfc.dcci.com

Rural Development Centre
http://www.une.edu.au/~trdc/RDC.HTM

Rural Economic and Community Development (USDA)
http://www.rurdev.usda.gov

Rural Europe Homepage
http://www.rural-europe.aeidl.be

Rural Information Center
http://www.nal.usda.gov/ric/

The Rural Policy Research Institute - RUPRI
http://www.rupri.org/

The Rural Research Centre at the Nova Scotia Agricultural College
http://www.nsac.ns.ca/nsac/rrc

Rural Sociological Society
http://www.lapop.lsu.edu/rss

Social Sciences Data Collection
http://ssdc.ucsd.edu/ssdc/8490.html

Statistics Canada
http://www.statcan.ca/start.html

Stat-USA Database Index
http://www.stat-usa.gov/BEN/databases.html

TVA Rural Studies Program
http://www.rural.org

UK Community Information Networks
http://panizzi.shef.ac.uk/community

United States Department of Agriculture (USDA)
http://www.usda.gov

USDA Economics and Statistics System
gopher://usda.mannlib.cornell.edu

W. K. Kellogg Collection of Rural Community Development Resources
http://www.unl.edu/kellogg/index.html

EDUCATION AND REFERENCE

AAPC on the Web
http://www.aacp.org

Biological, Agricultural, and Medical INFOMINE
http://lib-www.ucr.edu/bioag/
Contains links and information including library resources.

BUBL Information Service Web Server
http://bubl.ac.uk/

Educational Hotspots List from the Franklin Institute Science Museum
http://sln.fi.edu/tfi/jump.html

Educational Recycling Programs
http://www.umn.edu/nlhome/m010/recycle/educ.html

Guides to the Internet
gopher://gopher.merit.edu/

Hawaii Biological Survey
http://www.bishop.hawaii.org/bishop/HBS/hbs1.html
Surveys the Hawaiian Archipelago for flora and fauna (graphic intensive). It contains information about damselflies, fieldwork, databases, and publications.

Innovation
http://dc.smu.edu/dc/innovation.html

Madison Area Technical College
http://www.madison.tec.wi.us/

Midlink Magazine
http://longwood.cs.ucf.edu/~MidLink/
This is the electronic magazine for kids in the middle grades.

Online Internet Institute
http://prism.prs.k12.nj.us:70/0/WWW/OIIftf.html

Online Reference Works
http://www-cgi.cs.cmu.edu/web/references.html

Poster Net
http://pharminfo.com/poster/pnet_hp.html

The Saul High School for Agricultural Sciences
http://www.phila.k12.pa.us/schools/saul/

Summer on the Farm Video Series
http://www.ltc.com/farmvideo/
Details an educational video series for children about farming.

The Switchboard
http://www.switchboard.com

Useful Internet Sites
http://witloof.sjsu.edu/iwtools/webtools.html

Virtual Facts on File
http://www.refdesk.com/facts.html
This site contains links to online encyclopedias, dictionaries, thesauri, etc.

Web66: A Kids World Wide Web Project
http://web66.coled.umn.edu/

Webster's Dictionary
http://c.gp.cs.cmu.edu:5103/prog/webster?

Wisconsin School of Electronics
http://www.herzing.edu/madison/

Yahoo's List of Schools and Educational Sites on the WWW
http://www.yahoo.com/Education/

GOVERNMENT

Agriculture Marketing Service (USDA)
http://www.ams.usda.gov/

Bureau of Labor Statistics
http://stats.bls.gov/blshome.html

Congress.org
http://congress.org

Consumer Product Safety Commission
http://www.cpsc.gov

Department of Defense
http://www.dtic.mil/defenselink

Department of Education
http://www.ed.gov

Department of Health and Human Services
http://www.os.dhhs.gov

Department of Housing and Urban Development - HUD
http://www.hud.gov

Department of Interior and Bureau of Land Management
http://www.blm.gov

Department of Labor
http://www.dol.gov

Department of Transportation
http://www.dot.gov

Department of Treasury - IRS
http://www.irs.ustreas.gov

Department of Veterans Affairs
http://www.va.gov

Environmental Protection Agency
http://www.epa.gov

Federal Emergency Management Agency - FEMA
http://www.fema.gov

Federal Trade Commission
http://www.ftc.gov

Internet Addresses

FedWorld Government Job Listing
http://www.fedworld.gov/#jletr

Government and Politics Table of Contents
http://webcrawler.com/select/ref.govt.html

Government Information Sharing Project from OSU
http://govinfo.kerr.orst.edu

Government Resources from the University of Arkansas
http://www.uark.edu/world/subjects/government.html

House of Representatives
http://www.house.gov

Judicial Branch
http://www.whitehouse.gov/WH/html/judg-plain.html

Legislative Branch
http://www.whitehouse.gov/WH/html/legi.html

National Aeronautics & Space Administration - NASA
http://www.nasa.gov

National Agricultural Library
http://www.nalusda.gov

National Oceanic and Atmospheric Administration
http://www.noaa.gov

National Science Foundation
http://www.nsf.gov

Peace Corps
http://www.peacecorps.gov

Social Security Administration
http://www.ssa.gov

Tennessee Valley Authority - TVA
http://www.tva.gov

Thomas: Legislative Information on the Internet
http://thomas.loc.gov/

United States Department of Agriculture - USDA
http://www.usda.gov

U.S. Fish and Wildlife Service
http://www.fws.gov

U.S. Geological Survey
http://www.usgs.gov/USGSHome.html

U.S. Patent and Trademark Office
http://www.uspto.gov

U.S. Postal Service
http://www.usps.gov

The Whitehouse
http://www.whitehouse.gov

Whitehouse Offices and Agencies
http://www.whitehouse.gov/WH/EOP/html/EOP_org.html

World Wide Web Power Index of Government Sites
http://www.webcom.com/power/govt.html

Yahoo's Government List of Links
http://www.yahoo.com/Government

HEALTH

Atlanta Reproductive Health Centre
http://www.ivf.com/index.html

The Canadian Health Network
http://www.hwc.ca/links/english.html

CH.A.D.D. (Children and Adults with Attention Deficit Disorders)
http://www.chadd.org

Health Information List from Yahoo
http://www.yahoo.com/Health/

Health Hotlist from the Franklin Institute of Science Museum
http://sln.fi.edu/tfi/hotlists/health.html

Health Related Internet Resources by Subject
http://www.hsc.missouri.edu/main_ndx/health/subjind.html

Healthwise from Columbia University
http://www.columbia.edu/cu/healthwise

International Food Information Council
http://ificinfo.health.org

Kid's Health
http://www.dc.enews.com/magazines/healthykids/

Montreal Institute of Reproductive Medicine
http://www.infertility.ca/indexe.htm

National Clearinghouse for Alcohol and Drug Information
http://www.health.org/

National Health Information Center - NHIC
http://nhic-nt.health.org/

National Health Security Plan Table of Contents
http://sunsite.unc.edu/nhs/NHS-T-o-C.html

National Institute of Health
http://www.nih.gov

OncoLink from the University of Pennsylvania Cancer Resource Center
http://cancer.med.upenn.edu

RuralNet Rural Health Care
http://ruralnet.mu.wvnet.edu/

U.S. Department of Health and Human Services
http://www.os.dhhs.gov

West Virginia University Health Sciences Center
http://www.hsc.wvu.edu/main/

Women's Health Hot Line Home Page
http://www.soft-design.com/softinfo/womens-health.html

World Health Organization
http://www.who.ch

The World Wide Web Virtual Library: Biosciences - Medicine
http://www.ohsu.edu/cliniweb/wwwvl/

HISTORY

American History Educational Resources on the WWW
http://members.tripod.com/~catfan/hist.htm
College and high school students will find educational resources and reference material here for their studies in American history.

Florida Museum of Natural History
http://www.flmnh.ufl.edu/
Information on more than 10,000,000 specimens of mammals, birds, reptiles, etc.

The History of Technology
http://www.englib.cornell.edu/ice/lists/historytechnology/historytechnology.html
Contains links, essays and journals about the history of science and technology.

History Net—Where History Lives on the Web
http://www.thehistorynet.com/
Contains indexed articles, weekly features and links to history sites.

Illinois Natural History Survey
http://denr1.igis.uiuc.edu:70/
Provides information on topics such as biodiversity, economic entomology, aquatic ecology, and wildlife ecology, and includes online INHS databases, news and announcements, and geographic information system data and projects.

The Institute and Museum of History of Science (Florence, Italy)
http://galileo.imss.firenze.it/
Includes around 5,000 original items, divided into two fundamental categories: the apparatus and scientific instruments of the Medici, and the Lorenese collection of instrument and didactic and experimental devices.

Institute of Agricultural History and the Museum of Rural History
http://www.rdg.ac.uk/Instits/im/home.html
Includes an extensive collection of archives; a reference library; the Museum of English Rural Life; a photograph collection dating from the 19th century onwards; and the Bibliography of British and Irish Rural History.

The International Institute of Social History
http://www.iisg.nl/

National Center for History in the Schools
http://www.sscnet.ucla.edu/nchs/
Publishes the Guidelines of the National Standards for United States History, World History, and K-4 History.

Yahoo's History Directory
http://www.yahoo.com/Arts/Humanities/History/
Includes history links indexed by category.

HOME AND GARDEN

Algy's Herb Page
http://www.algy.com/herb/index.html

Arizona Poison Page
http://amber.medlib.arizona.edu/poison.html
This site contains lists of links to venomous animals, poisoning, toxicity, chemical hazards, and other poison control resources.

Backyard Compost Bins
http://www.envirolink.org/archives/recycle/0025.html

Childhood Lead Poisoning Prevention
http://TheArc.org/faqs/leadqa.html

Composting At Home
http://www.gvrd.bc.ca/waste/bro/swcomp1.html

Composting Information from Environment Canada Atlantic Region
http://atlenv.ns.doe.ca/udo/paydirt.html

Contech - Intelligent Animal Control
http://www.scatmat.com
This page offers a source of information on high tech pet and animal training/repelling products for vets, trainers, gardeners, farmers and pet owners.

Cornell University Fruit and Vegetable Information
http://www.cce.cornell.edu/topics/agriculture.html

Culinary Resources on the World Wide Web
http://www.gumbopages.com/food-www.html

Erma's Herbs
http://www.bizcafe.com/sspecial/ermaherb.html

Gardening Archive
http://www.lysator.liu.se/garden/index.html

Herb and Spice Stores
http://dinnercoop.cs.cmu.edu/dinnercoop/stores/spices.html

Herbs and Spices
http://www.teleport.com/~ronl/herbs/herbs.html

Hobbies and Crafts List from Yahoo
http://www.yahoo.com/Recreation/Hobbies_and_Crafts/

Home and Gardening List from Yahoo
http://www.yahoo.com/Recreation/Home_and_Garden/

Home Canning Guide
http://hammock.ifas.ufl.edu/txt/fairs/31520

Horticulture Solutions
http://www.ag.uiuc.edu/~robsond/solutions/horticulture/soils.html

Injury Control Resource Info Network
http://www.injurycontrol.com/icrin

Missouri Botanical Garden
gopher://gopher.mobot.org:70/1
Contains information on the topics of flora of North America, moss, and plants in bloom and allows you to use the library and search the index.

Poison Control Centre Database
http://vhp.nus.sg/PID/PCC/centre.html
This site contains a list of online poison control centers worldwide. To find U.S. sites, just scroll down the page.

Recipe List from Yahoo
http://www.yahoo.com/Entertainment/Food_and_Eating/Recipes/

Rot Web Home Composting Information Site
http://net.indra.com/~topsoil/Compost_Menu.html

Southern Living Online
http://www.pathfinder.com/@@Vs5u*wcAjQPBJsnq/sl/

Urban Agriculture Notes On Composting
http://www.cityfarmer.org/paulcomp66.html

Utah State University Extension Service's Garden Page
http://ext.usu.edu/yard/index.htm
This site is intended to help homeowners plan their outdoor living areas. Includes current publications and helpful links.

LAW

HIEROS GAMOS - The Comprehensive Legal Site
http://hg.org/journals.html
This is an EXTREMELY comprehensive law site. It includes indexes and links to almost all law related sites.

National Center for Agricultural Law Research and Information
http://law.uark.edu/arklaw/aglaw/
The NCALRI maintains an agricultural law library collection and also maintains current bibliographies on specific agricultural law topics.

Preserve/Net Law Service
http://www.preservenet.cornell.edu/law/plawmain.htm

The U.S. House of Representatives Internet Law Library
http://law.house.gov/95.htm

Welcome To Law.Net
http://law.net

NATURAL RESOURCES

Colorado Water Resources Research Institute
http://www.ColoState.EDU:80/Depts/CWRRI/
This page contains a link to the Colorado Water Newsletter, as well as links to CWRRI's research program and the National Institute for Water Resources.

Co-management of Aboriginal Resources
http://www.lib.uconn.edu/ArcticCircle/NatResources/comanagement.html
This site addresses the conflict related to natural resource management in development situations where resource management affects aboriginal communities.

Eco Web
http://ecosys.drdr.virginia.edu/EcoWeb.html
This is an environmental information resource page.

The Green Center
http://www.fuzzylu.com/greencenter/

Integrated Natural Resource Systems
http://lep.cl.msu.edu/msueimp/htdoc/modsr/03229570.html
From Michigan Agricultural Experiment Station, Michigan State University.

Integrating the Free Market Economic System in the Natural Economics System
http://rfweston.com/sd/free.htm

University of Connecticut College of Agriculture and Natural Resources
http://www.lib.uconn.edu/CANR/
Contains details about agricultural and resource economics, animal science, cooperative extension, experiment station, natural resources management ,etc.

Internet Addresses

University of Nebraska-Lincoln Institute of Agriculture and Natural Resource Database
http://www.ianr.unl.edu/cgi/home.pl
Information about the Institute of Agriculture and Natural Resources.

USDA Natural Resources Conservation Service
http://www.nrcs.usda.gov/
The Natural Resources Conservation Service is the federal agency that works with landowners on private lands to conserve natural resources.

Washington State University Center for Sustaining Agriculture and Natural Resources
http://csanr.wsu.edu/
This site contains information about sustainable agriculture and natural resources. It includes current features, links and a searchable database.

NEWS

ABC.com
http://www.abc.com

America's Job Bank
http://www.ajb.dni.us/

CBS News
http://www.cbsnews.com/

CNN News Interactive
http://cnn.com/

C-Span
http://www.c-span.org/

The Electronic Newstand
http://www.enews.com

FedWorld Government Job Listing
http://www.fedworld.gov/#jletr

Fox Online
http://www.fox.com

Internet Jobs and Career Resources from the University of Arkansas
http://www.uark.edu/world/subjects/jobs.html

JobHunt: On-Line Job Meta List
http://www.job-hunt.org/

Lycos Top News
http://www.topnews.com/

Media Online Yellow Pages
http://www.webcom.com/~nlnnet/yellowp.html
Contains lists of media resources on the Internet. Includes network, distributor, and show information links.

Minnesota Public Radio
http://www.mpr.org

National Geographic
http://www.nationalgeographic.com

National Public Radio
http://www.npr.org

NBC.com
http://www.nbc.com

The New York Times
http://www.nytimes.com

News List from Yahoo
http://www.yahoo.com/News/

News Page
http://www.newspage.com/
Provides daily trade news.

NEWS.Com
http://www.news.com

Newslink
http://www.newslink.org
Offers links to online newspapers, magazines, etc.

The Official Olympic Website
http://www.olympic.org/

People Magazine
http://www.pathfinder.com/people

The PointCast Network
http://www.pointcast.com

Public Broadcasting Service
http://www.pbs.org

Public Radio International
http://www.pri.org

Reuters
http://www.reuters.com

Southern Living Magazine Online
http://www.pathfinder.com/@@FiXhkAcA7ACN53uj/sl

Time
http://www.pathfinder.com/@@FiXhkAcA7ACN53uj/time/

USA Today
http://www.usatoday.com

Wall Street Journal
http://update.wsj.com/
To use this online site, you must subscribe.

The Washington Post
http://www.washingtonpost.com

WebOvision
http://www.webovision.com/cgibin/var/media/sd/index.html
Provides links to media around the world.

SEARCH ENGINES AND LISTS

The AccuFind Net Locator
http://nln.com/

Agricultural Colleges Online
http://w3.ag.uiuc.edu/AgColleges.Html
Contains a short list of universities that have agriculture colleges and departments providing WWW services.

AgriGator Index of Agricultural and Related Information
http://gnv.ifas.ufl.edu/WWW/AGATOR/HTM/AG.HT

AgriSurf
http://www.agrisurf.com

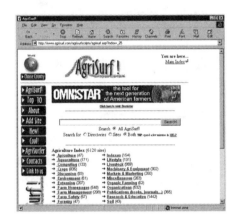

The All-in-One Search Engine
http://www.albany.net/allinone/

The AltaVista Search Engine
http://www.altavista.digital.com

AltaVista Suprise Cyberspace Jump
http://www.altavista.digital.com/cgi-bin/query?pg=s&what=web

The BigBook
http://www.bigbook.com

Bigfoot Search (e-mail addresses)
http://www.bigfoot.com

Caffeine Search (Canada Agriculture Search)
http://www.caffeine.ca

The Currency Converter by Olsen and Associates
http://www.olsen.ch/cgi-bin/exmenu

DejaNews
http://www.dejanews.com/

The Electric Library
http://www.elibrary.com

Excite Search
http://www.excite.com

Four11 Directory Services
http://www.four11.com

GTE Superpages
http://www.superpages.com

The HotBot Search Engine
http://www.hotbot.com

HotWired
http://www.hotwired.com

The Huge List of Links
http://thehugelist.com/links.html

The IBM InfoMarket
http://www.infomarket.ibm.com/

Infoseek
http://www.infoseek.com

Internet Addressbook
http://www.addressbook.com/

Internet Subject Guide from the University of Arkansas
http://www.uark.edu/world/subjects/

List of USENET FAQs
http://www.cis.ohio-state.edu/hypertext/faq/usenet/

Lycos a2z!
http://a2z.lycos.com

Lycos Search
http://www.lycos.com

The Magellan Search Engine
http://www.mckinley.com

Megalists of Psychology-Related Sites on the Web
http://www.gasou.edu/psychweb/resource/megalist.htm

The Open Text Index
http://index.opentext.net

Pointcast
http://www.pointcast.com

Rural Internet Access, Inc. Home Page
http://www.ruralnet.net

Search.com
http://www.search.com

Switchboard
http://www.switchboard.com

WebCrawler Searching
http://webcrawler.com

The WhoWhere? Search Engine (e-mail addresses)
http://www.whowhere.com

Wired
http://www.wired.com

World of Links
http://www.hooked.net/users/wcd/links.html

World Wide Web Virtual Library
http://www.w3.org/hypertext/DataSources/bySubject/Overview.html

Yahoo's List
http://www.yahoo.com

The Yellow Page
http://www.mcp.com

SPORTS AND RECREATION

CDNow
http://www.cdnow.com
Online music store.

City.Net North America
http://city.net/regions/north_america/

College Sports Network
http://www.xcscx.com/colsport/
Some sites are maintained by the college's athletic department and some by the students.

ESPNET Sports Zone
http://espnet.sportszone.com/

Golf.com
http://www.golf.com/

Golfweb
http://www.golfweb.com

GORP - Great Outdoor Recreation
http://www.gorp.com

High Fun, High Adventure, High Altitude
http://www.mountainzone.com

The National Park Service's ParkNet
http://www.nps.gov

Outdoor List from Yahoo
http://www.yahoo.com/Recreation/Outdoors/

Palo Alto Baylands Preserve
http://xymox.palo-alto.ca.us/av/baylands.html

Pennsylvania Farm Vacation Association
http://www.pafarmstay.com
Pennsylvania Farm Vacation Association is an organization of family farms across the state that offer overnight accomodations.

Pollstar
http://www.pollstar.com
Concert information.

Rec. Travel Library: Worldwide Travel and Tourism Information
http://www.remcan.ca/rec-travel/

A Sportsmans Dream - Wildlife Photography
http://www.onlinecol.com/sd/indexe.htm

Ticketmaster
http://www.ticketmaster.com
Reserve and purchase Ticketmaster tickets online!

Ultimate Band List
http://www.ubl.com
Huge list of bands and information.

Virtual Tourist I
http://www.vtourist.com/webmap/

Yahoo! Maps
http://www.vicinity.com/yahoo/

STATE INFORMATION & UNIVERSITY RESOURCES

Cooperative Extension System Information Servers by State
http://www.esusda.gov/statepartners/
This site is a service of the Cooperative State Research, Education, and Extension Service, and it provides a list of cooperative extension sites on the Internet by state.

Sites of Interest by State from Yahoo
http://www.yahoo.com/Regional/U_S_States/
Yahoo maintains a list of interesting sites for each state in the United States. The list includes links to business, community, education, entertainment, government, organizations, and travel information for each state.

State Search by NASIRE
http://www.nasire.org/ss/index.html
State Search is a service of the National Association of State Information Resource Executives and is designed to serve as a topical clearinghouse to state government information on the Internet.

Universities on the WWW from Yahoo
http://isl-garnet.uah.edu/Universities/
The list of universities on the WWW is maintained at the Intelligent Systems Laboratory, of the Center for Automation and Robotics, at the University of Alabama in Huntsville. It includes listings of all state universities and colleges on the Internet.

TAX INFORMATION

Department of Revenue
http://www.state.sd.us/state/executive/revenue/revenue.html
This site contains tax resources including information about tax forms and applications, individual taxes, and tax laws.

IRS - The Digital Daily
http://www.irs.ustreas.gov/prod/cover.html
This is a daily digital newspaper published by the IRS containing updated tax information and filing resources.

IRS Forms and Publications
http://www.irs.ustreas.gov/plain/

Secure Tax - Online Electronic Filing
http://www.securetax.com

Tax Forms and Instructions
http://www.fedworld.gov/ftp.htm

Taxes
http://www.exxnet.com/taxes.htm
Provides quick tax help information on tax law changes, plus tips on how to cut your tax bill.

Yahoo! - Business and Economy:Taxes
http://www.yahoo.com/Government/Taxes/
Contains Yahoo's list of links to tax tips and tax resources.

TELECOMMUNICATIONS

AppleGate Media
http://www.apgate.com/

Benton Foundation Communications Policy Project
http://www.cdinet.com/Benton/home.html

DataMasters—1997 Computer Industry Salary Survey
http://www.datamasters.com/survey.html
DataMasters specializes in supporting all areas of information systems.

Federal Communications Commission
http://www.fcc.gov

Framework for Global Electronic Commerce
http://www.iitf.nist.gov/eleccomm/glo_comm.htm

General Telecom Resources
http://www.contrib.andrew.cmu.edu/~blt/telecom_srcs.html

Government Technology: Electronic Commerce
http://www.govtech.net/1997/gt/apr/april1997-ecgrid/april1997-ecgrid.shtm

International Telecommunication Union - ITU
http://www.itu.int/

LearnNet From the Federal Communications Commission
http://www.fcc.gov/learnnet/
This is the FCC's informal education page.

PacBell's ISDN Home Page
http://www.pacbell.com/isdn/isdn_home.html

Qwest Communications
http://www.erinet.com/qwest/
Offers Web site design, desktop publishing and media and public relations.

Telecom A.M.
http://www.telecommunications.com/am/
This site covers telecommunication news issues.

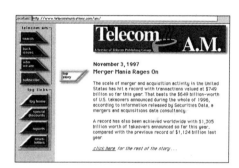

Telecom Information
http://www.fokus.gmd.de/nthp/
telecom-info/entry.html

Telecom Information Resources
http://www.spp.umich.edu/telecom/
telecom-info.html

(Tele)Communications Information Sources
http://www.telstra.com.au/info/communications.html
Lists a number of information sources in the field of communications and telecommunications.

Telecommuting, Teleworking, and Alternative Officing
http://www.gilgordon.com/

West Georgia Telecommunications Resource Directory
http://www.wgta.org

Yahoo's Telecommunications Information
http://www.yahoo.com/Business/corporations/telecommunications
This is Yahoo's list of telecommunications links.

TELEPHONE COMPANIES

AT&T
http://www.att.com

Excel
http://www.exceltel.com

MCI
http://www.mci.com

Sprint
http://www.sprint.com

TRAVEL

AAA
http://www.aaa.com

Best Western
http://www.bestwestern.com

City.Net
http://city.net

Days Inn
http://www.daysinn.com

Hampton Inn
http://www.hampton-inn.com

Hilton
http://www.hilton.com

Holiday Inn
http://www.holiday-inn.com

Howard Johnson
http://www.hojo.com

Hyatt
http://www.hyatt.com

International Guide to Bed & Breakfast Inns
http://www.ultranet.com/biz/inns

International Travelers Clinic
http://www.intmed.mcw.edu/ITC/Health.html

Knights Inn
http://www.knightsinns.com

Marriott
http://www.marriott.com

Olsen and Associates Currency Converter
http://www.oanda.com/site/ccc_intro.shtml

Quality Inn
http://www.qualityinn.com

Radisson
http://www.radisson.com

Ramada Inn
http://www.ramadahotels.com

Rec.Travel Library
http://www.travel-library.com

Red Roof Inn
http://www.redroof.com

Sheraton
http://www.sheraton.com

Super 8
http://www.super8motels.com

Travel.org
http://www.travel.org

TravelWeb
http://www.travelweb.com

Virtual Tourist
http://www.vtourist.com/webmap

Yahoo! Maps
http://maps.yahoo.com/yahoo

WEATHER
USA WEATHER

AccuWeather
http://www.AccuWeather.com

Air Force Weather Service
http://infsphere.safb.af.ml/users/aws/public_www/public

American Meteorolgical Society
http://atm.geo.nsf.gov/AMS

Aviation Weather Program
http://www.nws.noaa.gov/om/aviation.htm

Aviation Weather Center
http://www.awc-kc.noaa.gov

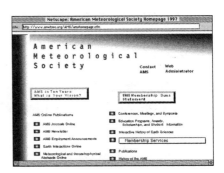

CNN Weather
http://www.cnn.com/WEATHER/index.html

Earthwatch Communications, Inc.
http://www.earthwatch.com

Emergency Managers Weather Information Network - EMWIN
http://www.nws.noaa.gov/oso/oso1/oso12/document/emwin.htm

▼ *Part II*

Federal Emergency Management Agency - FEMA
http://www.fema.gov

GOES Imagery
http://goeshp.wwb.noaa.gov

Interactive Weather Information Network - IWIN
http://iwin.nws.noaa.gov/iwin/graphicsversion/main.html

METAR/TAF
http://www.noaa.gov/oso/oso1/oso12/metar.html

Michigan State's Interactive Weather Browser
http://wxweb.msu.edu/weather

MSNBC / Intellicast Weather
http://www.intellicast.com/

National Climatic Data Center - NCEP
http://www.ncdc.noaa.gov

National Data Buoy Center
http://seaboard.ndbc.noaa.gov

National Operational Hydrologic Remote Sensing Center
http://www.nohrsc.nws.gov

National Radar Overview
http://www.intellicast.com/weather/usa/radar

National Weather Service Homepage
http://www.nws.noaa.gov

National Weather Service State Forecasts
http://iwin.nws.noaa.gov/iwin/textversion/states.html (text version)
http://iwin.nws.noaa.gov/iwin/iwdspg1.html (graphics version)

NCEP Enviromental Modeling Center
http://nic.fb4.noaa.gov:8000/

NCEP Hydro-Meteorological Prediction Center
http://www.ncep.noaa.gov/HPC

NCEP Marine Prediction Center
http://www.ncep.noaa.gov/MPC

NCEP Space Environment Center
http://www.sec.noaa.gov

NESDIS Home Page
http://ns.noaa.gov/NESDIS/NESDIS_Home.html

NOAA Coastal Ocean Program
http://hpccl.hpcc.noaa.gov/cop-home.html

NOAA Enviromental Research Laboratory - ERL
http://www.erl.noaa.gov

NOAA Home Page
http://www.noaa.gov

NOAA Public Affairs
http://www.noaa.gov/public-affairs

North Carolina Cooperative Extension Service - NCCES
http://www.ces.ncsu.edu/
Contains links to North Carolina ag weather forecasts.

Northeast Regional Climate Center
http://met-www.cit.cornell.edu/nrcc_home.html

NWS Climate Diagnostic Center - CDC
http://www.cdc.noaa.gov/

NWS Climate Prediction Center
http://www.nic.fb4.noaa.gov

NWS National Center For Enviromental Prediction - NCEP
http://nic.fb4.noaa.gov:8000/

NWS Public Affairs
http://www.nws.noaa.gov/pa

NWS Office Of Hydrology
http://www.nws.noaa.gov/oh/index.html

NWS Office Of Industrial Meteorology
http://www.nws.noaa.gov/im

NWS Office of Meteorology
http://www.nws.noaa.gov/OM/omhome.htm

NWS Storm Prediction Center
http://www.nssl.uoknor.edu/~spc

Old Farmer's Almanac
http://www.almanac.com

Real Time Weather Network
http://weathernet.com

Skywarn
http://www.skywarn.org/

Southeast Agricultural Weather Center
http://www.awis.auburn.edu

Southeast Regional Climate Center
http://www.sercc.dnr.state.sc.us

Southern Regional Climate Center
http://maestro.srcc.lsu.edu

Southwest Agricultural Weather Center
http://www.swami.tamu.edu

Spaceflight Meteorology
http://shuttle.nasa.gov/weather/smghome.html

The Storm Chasers Page
http://taiga.geog.niu.edu/chaser/chaser.html

Successful Farming's Agriculture Online Weather
http://www.agriculture.com/weather/weather.html
Subscription required.

TV Station and Network Weather Departments on the Web
http://www.tvweather.com/tv_dept.htm

TVWeather.Com
http://www.tvweather.com

University of Kentucky Agricultural Weather Center
http://wwwagwx.ca.uky.edu

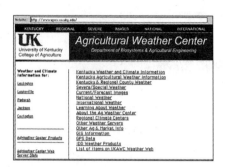

USA Today Weather
http://cgi.usatoday.com/weather/wfront.htm

The Weather Channel
http://www.weather.com

The Weather Lab
http://weatherlabs.com

REGIONAL WEATHER

Alaska Aviation Weather Unit
http://www.alaska.net/~aawu

Alaska Region HG
http://www.alaska.net/~nwsar
Includes forecasts from WFO Anchorage, WFO Fairbanks, and WFO Juneau.

Alaska River Forecast Center
http://www.alaska.net/~akrfc

Arkansas/Red River Basin RFC
http://info.abrfc.noaa.gov

Cal/Nev RFC, CA
http://nimbo.wrh.noaa.gov/Sacramento

Central Region HQ
http://www.crh.noaa.gov

Colorado Basin RFC, UT
http://www.cbrfc.gov/home.html

CWSU Jacksonville, FL
http://www.srh.noaa.gov/ftproot/zjx/index.html

Eastern Region HQ
http://www.noaa.gov/er/hq/index.htm

Goddard Space Flight Center
http://www.gsfc.nasa.gov/gsfc_homepage.html

Lower Mississippi RFC, Lake Charles, LA
http://www.srh.noaa.gov/ftproot/orn/html/default.html

Mid Atlantic RFC, PA
http://marfcws1.met.psu.edu

Missouri Basin RFC, MO
http://www.crh.noaa.gov/mbrfc

North Central RFC, MN
http://www.crh.noaa.gov/ncrfc/welcome.html

Northeast RFC, MA
http://www.nws.noaa.gov/er/nerfc

Northwest RFC, OR
http://www.nwrfc.noaa.gov

Ohio River RFC, OH
http://www.nws.noaa.gov/er/iln/ohrfc.html

Pacific Region HQ
http://www.nws.noaa.gov/pr/pacific.htm

Southeast RFC, Atlanta
http://www.srh.noaa.gov/ftproot/atr/html/main_p.htm

Southern Region HQ
http://www.srh.noaa.gov

West Gulf RFC, TX
http://www.srh.noaa.gov/wgrfc

Western Region HQ
http://www.wrh.noaa.gov

NATIONAL WEATHER SERVICE FIELD OFFICES ALPHABETICALLY BY STATE

WFO Birmingham, AL
http://www.acesag.auburn.edu/department/nws

WFO Mobile, AL
http://www.srh.noaa.gov/ftproot/mob/html/default.html

WFO Anchorage, AK
http://www.alaska.net/~nwsfoanc

WFO Little Rock, AR
http://www.srh.noaa.gov/FTPROOT/LZK/HTML/LZK.HTML

WFO Flagstaff, AZ
http://nimbo.wrh.noaa.gov/Flagstaff

WFO Phoenix, AZ
http://saguaro.la.asu.edu/nws

WFO Tucson, AZ
http://nimbo.wrh.noaa.gov/Tucson/twc.html

WFO Eureka, CA
http://www.northcoastweb.com/nws

WFO Los Angeles, CA
http://www.nwsla.noaa.gov

WFO Sacramento, CA
http://nimbo.wrh.noaa.gov/Sacramento

WFO San Diego, CA
http://nimbo.wrh.noaa.gov/Sandiego/nws.html

WFO San Francisco, CA
http://www.nws.mbay.net/home.html

WFO San Joaquin Valley (Fresno), CA
http://cnetech.cnetech.com/sjvwx

WFO Denver, CO
http://www.crh.noaa.gov/den

WFO Grand Junction, CO
http://www.crh.noaa.gov/gjt

WFO Pueblo, CO
http://www.crh.noaa.gov/pub/home.htm

WFO Jacksonville, FL
http://www.unf.edu/nws/jax/index.html

WFO Melbourne, FL
http://sunmlb.nws.fit.edu

WFO Miami, FL
http://www.srh.noaa.gov/FTPROOT/MFL/HTML/MFL.HTML

WFO Tallahassee, FL
http://www.nws.fsu.edu

WFO Tampa Bay, FL
http://www.marine.usf.edu/nws

WFO Atlanta, GA
http://www.srh.noaa.gov/ftproot/ffc

WFO Honolulu, HI
http://www.nws.noaa.gov/pr/os/nwsfo.htm

WFO Boise, ID
http://www.boi.noaa.gov

WFO Pocatello, ID
http://nimbo.wrh.noaa.gov/Pocatello/NWSPAGE2.HTM

WFO Central Illinois
http://www.crh.noaa.gov/ilx/ilxhome.htm

WFO Chicago, IL
http://taiga.geog.niu.edu/nwslot

WFO Des Moines, IA
http://www.crh.noaa.gov/dmx/index.html

WFO Quad Cities, IA
http://www.crh.noaa.gov/dvn/index.htm

WFO Indianapolis, IN
http://www.crh.noaa.gov/ind/start.htm

WFO Dodge City, KS
http://info.abrfc.noaa.gov/ddcdocs/nwsoddc.html

WFO Goodland, KS
http://www.crh.noaa.gov/gld/welcome.html

WFO Topeka, KS
http://www.crh.noaa.gov/top

WFO Wichita, KS
http://www.crh.noaa.gov/ict/index.html

WFO Louisville, KY
http://www.crh.noaa.gov/lmk/welcome.htm

WFO Paducah, KY
http://www.crh.noaa.gov/pah

WFO Lake Charles, LA
http://www.srh.noaa.gov/FTPROOT/LCH/HTML/LCH.HTML

WFO New Orleans, LA
http://www.srh.noaa.gov/FTPROOT/LIX/HTML/default.HTM

WFO Shreveport, LA
http://www.srh.noaa.gov/ftproot/shv/html/homepg.htm

WFO Boston, MA
http://www.nws.noaa.gov/er/box

WFO Gray (Portland), ME
http://www.seis.com/~nws/mainpgs2.html

WFO Detroit, MI
http://www.crh.noaa.gov/dtx/start.htm

WFO Grand Rapids, MI
http://www.crh.noaa.gov/grr/index.html

WFO Marquette, MI
http://www.crh.noaa.gov/mqt/mqt.htm

WFO N C Lower, MI
http://www.crh.noaa.gov/apx/homepage.htm

WFO Minneapolis, MN
http://www.crh.noaa.gov/mpx/mpx.html

WFO Duluth, MN
http://www.crh.noaa.gov/dlh/duluth.htm

WFO Kansas City, MO
http://www.crh.noaa.gov/eax/eax.htm

WFO Springfield, MO
http://www.crh.noaa.gov/sgf/sgf1.htm

WFO St. Louis, MO
http://www.crh.noaa.gov/lsx/lsx.htm

WFO Jackson, MS
http://www.jannws.state.ms.us/

WFO Hastings, NE
http:///www.crh.noaa.gov/gig/gidhome.htm

WFO North Platte, NE
http://www.crh.noaa.gov/lbf/lbf.home.html

WFO Omaha, NE
http://www.crh.noaa.gov/oax/front.html

WFO Elko, NV
http://nimbo.wrh.noaa.gov/Elko

WFO Las Vegas, NV
http://nimbo.wrh.noaa.gov/Lasvegas/

WFO Reno, NV
http://nimbo.wrh.noaa.gov/Reno

WFO Albuquerque, NM
http://www.srh.noaa.gov/FTPROOT/ABQ/HTML/ABQ.HTML

WFO Albany, NY
http://nwsfo.atmos.albany.edu/www/wx.html

WFO Binghamton, NY
http://sac.wbgm.noaa.gov

WFO Buffalo, NY
http://www.wbuf.noaa.gov/

WFO New York City (Brookhaven), NY
http://sun20.ccd.bnl.gov/~nws/

WFO Newport (Morehead City), NC
http://www.nws.noaa.gov/er/mhx/index.htm

WFO Raleigh, NC
http://www.nws.gov/er/rah

WFO Wilmington, NC
http://nwsilm.wilmington.net

WFO Bismarck, ND
http://www.crh.noaa.gov/bis/nwsbis.htm

WFO Eastern (Grand Forks), ND
http://www.crh.noaa.gov/fgf/index.html

WFO Cleveland, OH
http://www.csuohio.edu/nws

WFO Wilmington, OH
http://www.nws.noaa.gov/er/iln/iln.htm

WFO Norman, OK
http://www.nssl.uoknor.edu/~nws

WFO Tulsa, OK
http://info.abrfc.noaa.gov/wfodocs/wfotulsa.html

WFO Medford, OR
http://nimbo.wrh.noaa.gov/Medford/index.html

WFO Portland, OR
http://nimbo.wrh.noaa.gov/Portland

WFO Philadelphia, PA
http://www.nws.noaa.gov/er/phi

WFO State College, PA
http://bookend.met.psu.edu/webpages/homepage.html

WFO Columbia, SC
http://www.nws.noaa.gov/er/cae

WFO Charleston, SC
http://wch.csc.noaa.gov

WFO Aberdeen, SD
http://www.crh.noaa.gov/abr/nwsabr.htm

WFO Rapid City, SD
http://www.crh.noaa.gov/unr/index.htm

WFO Sioux Falls, SD
http://www.crh.noaa.gov/fsd/homepage.htm

WFO Knoxville, TN
http://www.srh.noaa.gov/FTPROOT/MRX/HTML/MRX.HTM

WFO Memphis, TN
http://www.srh.noaa.gov/ftproot/meg/html/default.html

WFO Nashville, TN
http://www.nws.noaa.gov/sr/ohx/weather.htm

WFO Amarillo, TX
http://info.abrfc.noaa.gov/amadocs/ama.html

WFO Brownsville, TX
http://www.srh.noaa.gov/FTPROOT/BRO/HTML/BRO.HTML

WFO Corpus Christi, TX
http://www.srh.noaa.gov/crp

WFO El Paso, TX
http://nwselp.epcc.edu

WFO Fort Worth, TX
http://www.srh.noaa.gov/FTPROOT/FWD/HTML/FWD.HTML

WFO Houston, TX
http://www.srh.noaa.gov/ftproot/hgx/html/homey1.htm

WFO Lubbock, TX
http://dryline.nws.noaa.gov

WFO Midland, TX
http://www.srh.noaa.gov/FTPROOT/MAF

WFO San Angelo, TX
http://www.srh.noaa.gov/FTPROOT/SJT

WFO San Antonio, TX
http://www.srh.noaa.gov/ewx

West Gulf RFC, TX
http://www.srh.noaa.gov/wgrfc

WFO Salt Lake City, UT
http://nimbo.wrh.noaa.gov/Saltlake/slc.noaa.html

WFO Blacksburg, VA
http://www.bev.net/weather

WFO Sterling, VA
http://www.nws.noaa.gov/er/lwx

WFO Wakefield, VA
http://www.gc.net/wxman1

WFO Burlington, VT
http://www.nws.noaa.gov/er/btv

WFO Seattle, WA
http://www.seawfo.noaa.gov

WFO Spokane, WA
http://nimbo.wrh.noaa.gov/Spokane/index.html

WFO Charleston, WV
http://yoopr.wrlx.noaa.gov

WFO Green Bay, WI
http://www.crh.noaa.gov/grb

WFO La Crosse, WI
http://www.crh.noaa.gov/arx/index.html

WFO Milwaukee, WI
http://www.crh.noaa.gov/mkx/welcome.htm

WFO Cheyenne, WY
http://www.crh.noaa.gov/cys/cyshome.htm

WFO Riverton, WY
http://riw.weather.wyoming.com/riwindex.htm

CANADIAN WEATHER

Environment Canada's Green Lane Weather Site
http://www.doe.ca/weather_e.html

Weather Farm and Country Ag Publishing Company
http://www.agpub.on.ca/weather.htm

INTERNATIONAL WEATHER

WFO Barrigada, Guam
http://www.nws.noaa.gov/pr/guam/p1.htm

WFO San Juan, PR
http://www.upr.clu.edu/nws

APPENDIX A
INTERNET SERVICE PROVIDERS

There are thousands of Internet Service Providers and online services. The following list obviously includes only a few of the better known ones. Note that most ISPs and online services bill your credit card and many won't direct bill or take checks. Before calling one of these national ISPs, you might want to check out any local ones. Local ISPs usually provide better customer service and more "hand holding" to help you get online. To find out if there are any local ISPs in your area, Look in the Yellow Pages under Computers or Internet, talk with Internet users in your community or check with local schools and libraries.

AMERICA ONLINE
Web: http://www.aol.com
Phone: 1-800-827-6364
Services: 28.8k dial-up, e-mail, personal Web pages and content you won't find anywhere else on the Internet (more oriented toward consumer needs than CompuServe)
Pricing: $19.95 unlimited access with additional pricing plans available
Areas Served: World
List of Access Numbers: The free software from America Online automatically determines this for you. However, this automatic system is not foolproof. See Appendix B.

Notes: Free software, very easy to use, but may require a long-distance call from many rural areas.

AT&T WORLDNET SERVICE
Web: http://www.att.com/worldnet
Phone: 1-800-WORLDNET
Services: 28.8k dial-up, e-mail, personal Web page
Pricing: $19.95 unlimited access for anyone, or AT&T long distance customers can choose $4.95/mo. for 5 hours (non-customers, $4.95 for 3 hours), $2.50 per additional hour
Personal Web Page: $1.95/mo. for 2 megabytes of space
Areas Served: Continental United Areas, Hawaii, Puerto Rico and the U.S. Virgin Islands

List of Access Numbers: http://www.att.com/worldnet/wis/faqs/access_states.html

Notes: Service is targeted toward AT&T long distance customers. Users who are dissatisfied and cancel can have their e-mail forwarded to their new provider for 60 days without charge. Free software included.

CAMPUS MCI

Web: http://www.campus.mci.net
Phone: 1-800-307-4481
Services: 28.8k dial-up, e-mail
Pricing: $12.00/mo. for 60 hours of usage
Area Codes Served: 213, 312, 316, 334, 406, 407, 408, 410, 423, 502, 505, 602, 606, 615, 619, 704, 715, 801, 803, 864, 901, 910, 916, 919, 954

Notes: This service is targeted toward students and faculty of participating universities, but is generally made available to anyone who is even loosely affiliated with a university. Free software. 800 access is available at a surcharge.

COMPUSERVE

Web: http://www.compuserve.com
Phone: 1-800-848-8199
Services: 28.8k dial-up, e-mail, and content you won't find anywhere else on the Internet (more oriented toward business needs than AOL)
Pricing: $19.95 unlimited access with additional pricing plans available
Area Served: World
List of Access Numbers: The free software from CompuServe automatically determines this for you.

Notes: Free software, very easy to use.

CONCENTRIC NETWORK
Web: http://www.concentric.net
Phone: 1-800-939-4262
Services: 28.8k dial-up, e-mail, personal Web site
Pricing: $19.95/mo. unlimited use, or $7.95/mo. for 5 hours, $1.95 per additional hour
Area Served: Continental United States

MCI INTERNET
Web: http://www.mci.com
Phone: 1-800-550-0927
Services: 28.8k dial-up, e-mail
Pricing: $19.95/mo. unlimited access, or $3.00 for 3 hours ($1.80 per additional hour)
Area Served: Continental U.S. and Hawaii

Notes: Free software.

MINDSPRING ENTERPRISES INC.
Web: http://www.mindspring.com
Phone: 1-800-719-4332
Services: 28.8k dial-up, e-mail
Pricing: $19.95/mo. unlimited access, or $14.95/mo. for 20 hours ($1 per additional hour)
Area Served: Continental United States

Notes: Access numbers and an online sign-up form are available on their home page (above).

NETCOM INTERNET SERVICES
Web: http://www.netcom.com
Phone: 1-800-NETCOM1
Services: 28.8k dial-up, e-mail
Pricing: $19.95 unlimited access
Area Served: Continental United States

APPENDIX B
RURAL INTERNET ACCESS

If you have a telephone line, you can definitely access the Internet. All you have to do is have your computer call the Internet Service Provider's access line. If this call is local, all you pay for Internet service is the ISP's monthly (or hourly) rate. But if this call is long distance, you will have to pay the ISP's charges plus the cost of the long-distance call. Since people usually spend more time online than they would on a regular phone call, the long-distance charges can quickly exceed the cost of the monthly Internet service itself.

For example, say that your ISP charges $19.95 per month for unlimited Internet access. If you can connect to your ISP with a local call, then $19.95 per month is all you pay. But if your ISP doesn't have a local access number that you can call for free, you'll have to make a long-distance call to get to your ISP. Since most people pay anywhere from 10 to 20 cents per minute (or more) for long-distance service (depending on the particular long-distance plan and time of day), these charges can make rural use of the Internet unaffordable. If you are paying, say, 20 cents per minute for long distance and use the Internet an average of just over three minutes per day, your long-distance bill will be about $20 per month, making your total Internet costs around $40 per month (assuming your ISP charges about $20). Three minutes per day is not very much time to spend on the Internet. It'll take you about one minute just to get connected and log on, so you would really have only about two minutes to spend "surfing" the Web. If you use the Internet a lot, your long-distance bill could be several times the cost of the Internet service itself.

The key to using the Internet from rural areas is to find a local access number. If one is not available, there are several steps you can take to reduce the cost of accessing the Internet (see below). However, none of them is as good as using a local access number. Since we wrote the first edition of the *Farmer's Guide*, local access to the Internet from rural areas has improved dramatically. Many more rural areas now have local Internet access numbers, and more will get local access. However, some areas still don't have local access. Where you live may be one of them, but make absolutely certain that you really can't dial a local number before accessing the Internet using long distance. Many people who thought they didn't have a local number were surprised to learn (after some checking) that they really did have local access.

Finding out if there's a local access number you can call for free is not too hard, but it may involve some detective work on your part. The national Internet providers whose names you probably already recognize (like Netcom, America Online, and CompuServe) know the names of the cities where their access lines are actually located. But they usually don't know about all the telephone exchanges in a given area, and probably won't know which exchanges you can make a free local call to.

For example, America Online (AOL) serves Marion, Arkansas, but they don't know they do. The area code for Arkansas is 501. If you tell AOL's computer that your area code is 501, it will give you a list of local numbers in the 501 area code you can call for free. In this example, AOL would list Jonesboro, Little Rock, and other cities with the 501 area code, but they are all long distance from Marion. However, for Marion residents, Memphis, Tennessee, is a free local call. But the area code for Memphis is 901, so Memphis wouldn't show up on a list of numbers for the 501 area code. Thus, Marion residents can connect to AOL through a "local" access number in Memphis. In fact, you don't even have to live in Marion—you only need to have a Marion phone number. Thus, a Crittenden County, Arkansas, farmer who lives near enough to town to have a Marion phone number can call AOL's Memphis access line, and it's still a free local call. The point is that if one of the large Internet providers has an access line in a nearby city that you can call for free, chances are you're the one who has to know. More often than not, they won't know.

Of course, there may be a small local ISP who has recently begun offering Internet access in your area and has local lines that you can call for free. But you've probably never heard of them and may not even know that they exist. Worse, since they're new, they probably have very little money to spend on advertising, so you may have to look hard to find them.

Internet Service Providers tend to have local numbers in cities—that's where most of their customers live. Whether or not a city has a local access line depends to a large degree on its population. There is no hard and fast rule, but most cities with populations of 50,000 or more usually have at least one ISP with a local access line, probably more. If you can make a free local call to a nearby city with a population of around 50,000, you should be able to find an ISP with a local access line there—and perhaps you'll find several to choose from.

Some smaller cities with populations of around 25,000 may also have local access to an ISP. And although the chances of having local access diminish as population decreases, in some cases even very small towns may have a local access line because of a college or government facility there.

The only way you'll know if you have local access is to do some research. The first step is to look in the front section of the White Pages of your telephone book for a list of all local telephone exchanges you can call for free. Don't rely on your memory; telecommunications regulations are changing, and your options may be greater today than even just a year or two ago. So check the current list of local exchanges, and see if any of the exchanges are in cities that might be large enough to have a local access number. If you find one, get a list of online services (see Appendix A) and call them to see if they have local access lines in that city. Be sure to tell them you are in the "large city" they probably serve, not in the small town where you really live, because they may not know of your town.

If none of the cities you can call for free has a local access line, then check to see if perhaps a small local ISP has established service in your area. They may be hard to find, so you'll have to look carefully. Here are some places to start:

WORD OF MOUTH

Check with other computer users in your area and see which ISPs they use and how they connect. Also, check with local schools, public libraries, nearby colleges, computer stores, and anyone else you think might have Internet access. Among farmers we talked to, the "word of mouth" method was often the most productive.

LOCAL NEWSPAPERS

Read the classified ads, particularly in the Computers or Electronic Equipment sections. Be sure to check the newspapers of all the towns to which you can make free calls, not just the town in which you live. If the newspaper has a reporter who writes about computers, give the reporter a call and ask about your local access options.

PHONE BOOK (YELLOW PAGES)

Look in the Yellow Pages of the phone book under Computers, Software and Services, Internet (or similar titles). But keep in mind that many ISPs may be too new or too cost-conscious to have Yellow Page ads yet. Check the Yellow Pages in the phone books of all of the cities you can call for free. If you don't have the phone book from another nearby city, borrow one from a friend who lives there, or check with the library.

LOCAL TELEPHONE COMPANY

Most if not all of the "Baby Bells" currently offer Internet access to at least part of their service areas. Many of the smaller, independent rural telephone companies and co-ops also offer access, so check with your local telephone company. Ironically, the smaller your phone company is, the more likely it is to offer Internet service.

RURAL ELECTRIC COOPERATIVES

More and more rural electric cooperatives are beginning to provide Internet access. Many that don't are "thinking about it," but some are unsure of the market potential. Inquiries from prospective customers might help them with their decision.

THE INTERNET ITSELF

It should come as no surprise that the Internet, which is a great source of information about nearly everything, is also a good source of information about the Internet itself. If you can get access to the Internet (using a friend's computer), you can search for ISPs by state, by area code, or by the name of the town(s) you can call for free. There are even large lists of ISPs you can check to see if any have service where you live. One good list is **thelist.internet.com**.

If you cannot find an Internet provider with a local access number that you can call for free, then try the following alternative, low-cost ways to access and use the Internet from rural areas.

GLASGOW GETS WIRED

A small municipally owned electric cooperative in Glasgow, Kentucky, (about 100 miles south of Louisville) is doing something that few telephone or cable companies in the country can match. They've wired the entire town of about 14,000 residents with a high-speed computer network—their own Information Superhighway.

In the late 1980's, the Glasgow Electric Plant Board (EPB), which gets its power from TVA, set up a two-way digital communications network in order to monitor electrical usage. But reading electric meters requires only a fraction of the network's total bandwidth. With all that leftover bandwidth available,

Glasgow EPB decided to go into the telecommunications business. First, they started with cable TV. But recently, Glasgow EPB began offering full-service Internet connections to their residential and business customers.

And what a connection! Because of the bandwidth available, Glasgow EPB's customers can access the Internet at speeds more than 60 times faster than using phone lines—even with a 28,800 bps modem.

What does all this speed cost? Surprisingly little: only $11.95 per month for unlimited access (plus a $9.95 per month rental charge for a special interface hardware that gets installed in the customer's PC).

MAKING A LOW-COST INTERNET CONNECTION

The key to low-cost Internet access is finding even one Internet Access Provider (or online service) that you can call locally for free. You may not be able to find one, but the cost savings are so dramatic that you owe it to yourself to try very hard. And we strongly encourage you to keep looking—every few months or so. Providing Internet access is a rapidly growing industry, and if you keep looking, something may turn up.

If you simply cannot find a local access number, you may have no choice but to use long distance to reach an ISP. However, there are still several things you can do to reduce your access costs, even if you're dialing long distance, and we'll go over each one.

Some of the suggestions for reducing costs are based on limiting the amount of time you would normally spend online—working faster and smarter. If you adhere to these rules too drastically, they can take much of the fun out of being on the Internet. A good compromise would be to set an amount you intend to spend each month, and operate prudently (using cost-reducing practices that have little or no effect on your Internet "experience"). Then you can use the money you saved to indulge in some pleasure trips along the Information Superhighway.

We've arranged this information in the form of options you should try or consider as ways to reduce your access costs. These options are roughly presented in the order of increasing cost, increasing complexity, and increased work on your part.

OPTION 1—EXPANDED CALLING ZONE

(Approximately 1 to 5 cents per minute)

OK. You're positive you can't make a free call to a local access number. If, however, there is a nearby city that has an Internet access number, you may be able to make that number a local call, even if that city is currently long distance.

Some phone companies offer an expanded calling zone (or a similar service) that lets you increase the size of your local calling area by paying an extra charge. The charges for expanded calling plans vary. Typically, they run 1 to 5 cents per minute, depending on the size of the expanded calling area you've chosen and the time of day. But this is less than the lowest-cost long-distance plan you can find.

Whether or not this option is available to you depends on the telephone regulations of your state and your phone company's policies. In some states, like Arkansas, Alabama, and Tennessee, customers have the option of expanded calling zones, while customers in other states, like Iowa, do not. The only way you'll know if you can do this is to check in the White Pages of your phone book (under Optional Calling Plans) or call your local phone company. As if 50 different state regulations were not enough confusion, each phone company has a different name for this service. Bell South calls its plan RegionServ, while Southwestern Bell uses the name Circle Saver.

Although you will end up paying an extra per-minute charge for this service, the cost is less than what you would pay for interstate long distance. (If it isn't, you shouldn't use it.) For example, in Tennessee, under Bell South's RegionServ plan, you would pay by the minute depending on how far you are from the city you are calling. If you are between 17 and 30 miles away, the cost is 5 cents per minute. If you are between 31 and 40 miles away, you pay 10 cents per minute. Furthermore, these rates are cut in half after 8 p.m. on weekdays and all day on weekends and holidays. So, if you confined your Internet activity to those times, you'd be paying either 2 1/2 or 5 cents per minute, which is less than half the cost of most low-cost interstate long-distance plans. In some cases, it's only one-fourth as much.

Expanded calling zones are an option for you if:
(1) your state telephone regulations allow them;
(2) your phone company offers them;
(3) you live within the required distance from the city you want to call;

(4) the city you are calling has a local Internet access number; and (5) the access line you will be connecting to is fast (28,800 bps or faster). If the access line is slow (say 14,400 bps), it is possible that you could come out ahead paying long-distance rates and using a faster (28,800 bps) connection.

This last item—connection speed— is important, but it is not immediately obvious. Not all local access lines connect at the same speed. Some can be very slow. Smaller cities tend to have the slowest connection speeds—some as low as 9,600 bps—because older, slower modems are cheaper for ISPs to buy. But their savings could cost you. A cheap call to a slow Internet connection can be more expensive than a long-distance call to a faster (28,800 bps) connection.

As a rule of thumb, an expanded calling zone is usually the best deal (compared with long-distance or toll-free charges) if the speed is 28,800 bps (or faster) and the expanded calling zone rate is less than 8 cents per minute.

FOREIGN EXCHANGE

An alternative to the expanded calling zone is called a foreign exchange or FX line. (Some people refer to them as leased lines.) Foreign exchange lines are available in some areas, but not all. And although they can be very expensive, it is also possible for an FX line to be a lower-cost option than toll-free or long distance.

Foreign exchange works like a calling zone, except that instead of paying by the minute, you pay a flat rate every month. (You may have to pay a hefty installation charge, too.) With an FX line, the phone in your home works like it's really located in another city. And, unlike expanded calling zones which usually apply to outgoing calls only, an FX line works both ways, so you can get "free" incoming calls from that city, too. Although an FX line may be expensive, it can be cheaper in the long run, especially when you take into account that it will eliminate long-distance calls you already make to that nearby city. Thus, when determining whether an FX line makes sense for you, include the cost of the calls you currently make to that city along with your anticipated Internet usage.

Because few people need FX lines, information about them usually isn't found in the White Pages of the phone book. Instead, you'll need to call the phone company's business office to inquire about availability and pricing.

OPTION 2—TOLL-FREE ACCESS LINE

(Approximately 8 to 10 cents per minute)

The next-to-least-expensive option is to use a toll-free access line. Toll-free calls are not really free—someone has to pay for them, and that will be you. While you may not be accustomed to paying for toll-free calls, most online services and ISPs levy an extra surcharge for using their toll-free access lines. Even so, toll-free surcharges can be cheaper than interstate long distance rates, because the online services and large national ISPs can buy long distance "in bulk" and get better rates than you and I.

The current "going rate" for "toll-free" access is about 10 cents per minute. However, be sure to check prices yourself when you are ready to buy. They change constantly.

Some ISPs charge higher than long-distance rates for toll-free access. One national ISP charges 15 cents per minute if you use their toll-free line. They're not trying to rip you off either. They provide the toll-free service as a convenience to business customers who may have trouble dialing in through hotel switchboards or while on the road. Consequently, don't assume that a toll-free surcharge will be cheaper than what you would pay for regular long-distance service.

OPTION 3—INTERSTATE LONG DISTANCE

(Approximately 10–12 cents per minute)

The major interstate long-distance carriers (AT&T, MCI, and Sprint) all have discount plans with night and/or weekend rates of around 10 to 12 cents per minute. Of course, most people don't take advantage of them, but you can—more about that later.

If you are willing to pay around 10 to 12 cents per minute, you can use a long-distance carrier to reach almost any ISP in the nation. Since you can choose virtually any ISP, pick one with a fast 28,800 bps (or faster) connection that won't be busy and slow during the times you want to be on the Internet. Hourly fees and customer service are also a consideration. If you are using long distance, you can afford to be choosy and pick the best ISP around.

Because of funny telephone regulations, making a long-distance call within your own state usually costs more than making a long-distance call clear across the country—in some cases it's double! Actually, it's a little more complicated than that. Telephone regulations vary state by state, but in general, each state is divided into local calling zones called LATAs. Your local phone company handles long-distance calls within the LATA, while long-distance companies like AT&T, MCI, and Sprint handle calls between LATAs and between states. A "long-distance" call within a LATA (or within a state) can cost 20 cents per minute or more (even at night), but between LATAs and between states, the costs can be half that. For this reason, you don't want to just plug in a modem and call the ISP closest to you, or you could easily end up paying 20 cents per minute or more. Instead, choose an access number outside your LATA or outside your state.

National online services are not very helpful in this regard. When you first sign up with an online service, you are typically asked to enter your area code. The online service then shows you a list of local access numbers within that area code. If none of those numbers is truly local, then some may be in the same LATA or state, which is usually the most expensive "long-distance" call you can make.

PICKING A LONG DISTANCE PLAN

According to various surveys, something like two-thirds of American consumers don't take advantage of the discount rate plans offered by the various long-distance companies. Today, you can choose among several long-distance carriers, and business is very competitive. Both Sprint and MCI have plans that can reduce your night and weekend long-distance costs to around 10 to 11 cents per minute. However, even venerable AT&T has discount plans, too. So you may not even need to switch long-distance carriers in order to get a better deal. Call your existing carrier and get information about their reduced rate plans. Some long-distance plans are based on how much you spend each month, so before shopping for a new plan, estimate about how much your total long-distance bill will be, including your Internet activity.

Your local phone company will probably charge a small ($5.00) fee each time you switch long-distance companies—but sometimes, the long-distance company will pay this fee for you! Most parts of the U.S are served by the long-distance carriers listed on the next page. Toll-free numbers for long-distance carriers that serve your area can also be found in the White Pages of your phone book.

Here are the codes and toll-free numbers for several major long-distance carriers:

Carrier	Toll-free Number
AT&T	(800) 222-0300
LCI	(800) 524-4685
LDDS	(800) 275-0100
MCI	(800) 444-3333
Sprint	(800) 746-3767

Billing Intervals: Long-distance carriers not only vary in what they charge, they can also vary in how they charge. Some long-distance carriers charge in 1 minute intervals while others charge in shorter intervals, some as small as 6 seconds. While this difference may sound trivial, it's not, especially if you make a lot of short calls—and you will be if you're sending and receiving e-mail and doing other work offline. (See page B-15: *Operating Practices That Can Save Money*.)

If your carrier charges in 1-minute intervals, then a call of 70 seconds will cost the same as a 2-minute call. A call of 2 minutes and 1 second is charged as a 3-minute call, and so on. On the other hand, if your carrier bases its charges on a shorter interval—such as 6 seconds—then for a 61-second call you would be billed for 66 seconds, not 2 minutes. For calls of long duration, like 30 minutes, the differences in billing intervals doesn't much matter. But one of the ways to cut costs is to work "off line," making a lot of quick calls of short duration. In that case, the difference in how they charge you can quickly add up. The big three (AT&T, MCI, and Sprint) tend to charge in 1-minute intervals. The lesser known long-distance carriers, like Allnet, LCI and LDDS, bill in 30-second or 6-second intervals (and have low weekend and night rates, too).

As this is being written, telephone regulations are undergoing tremendous change. Congress recently passed sweeping telecommunications regulations, and soon cable companies and long-distance companies may be offering local phone service, while local phone companies may be offering long distance. Keep abreast of these changes, and check rates often to ensure that you are on the best plan according to your use and are getting the lowest rate.

INTERNET CONNECTION ALTERNATIVES

If you can't locate a commercial Internet Service Provider you can call for free, you should check around to see if your area is served by a FreeNet or a Bulletin Board System (BBS).

FREENETS

In many rural areas, the FreeNet serves as a local Internet Service Provider. A FreeNet works a little like an agricultural or electric co-op—it's usually a community-based provider set up by a college or other nonprofit organization and run by volunteers. FreeNets provide Internet service in areas that otherwise wouldn't have local access. Usually, they provide Internet service for free (hence the name), but most charge a "membership fee," and some impose small charges to use certain Internet services. However, if you are looking for very low-cost access and don't want to make long-distance calls, you'll find that FreeNets can provide good service at bargain prices.

As you would expect, some FreeNets are well-run, professional services that are indistinguishable from their commercial rivals (or better). Others are—well, free.

The kinds of services that FreeNets can provide will often depend on the type of connection they have to the Internet. To keep costs down, many FreeNets are not permanently connected to the Internet. Instead, they receive their Internet feeds—e-mail, mailing lists, and newsgroups—in batches periodically throughout the day, or even once each night. By contrast, an online service like AOL or a full-service ISP like Netcom is always connected to the Internet, so your e-mail is delivered (to them) almost instantly after it is sent.

For most people, it doesn't matter whether or not their ISP is connected to the Internet all the time. After all, most of the time your e-mail is just sitting in your mail box waiting for you to log on and get it. So these intermittent Internet feeds are not a problem. But a FreeNet that connects to the Internet only occasionally can't provide access to the Web. You need a permanent, high-speed Internet connection for that, and that is expensive. As a result, many FreeNets (and their customers) don't feel the need to pay those prices just to use the Web and are quite happy just getting text-based services like e-mail and newsgroups. More and more FreeNets are connected to the Internet 24 hours a day, providing a full range of services, including access to the Web. Word of mouth is probably the best way to find out if there's one in your area—check with other users, schools, libraries and computer stores.

FREENET SOFTWARE

Some FreeNets use the same kind of software commercial ISPs use, while others are basically BBSs (see below) that use very basic communications software (the kind you get free with most modems). Many FreeNets will supply you with software—or at least, they'll help you get it. More importantly, since FreeNets are run by volunteers who want to help people get access to the Internet, they usually provide a very high level of service.

Besides being cheaper, FreeNets offer several advantages over commercial ISPs. Since they are community-based, FreeNets have more information of local interest than national online services do. For example, schools use them to assist students with homework assignments, and more and more local governments are using FreeNets to provide information to their citizens. Hospitals and other community organizations are using them too.

Perhaps the best thing about FreeNets is their excellent training programs to help local residents get on and use the Internet effectively. Even if you can't make a local call to a FreeNet, you may want to sign up just for their Internet lessons.

BULLETIN BOARD SYSTEMS

Further down the information food chain are the Bulletin Board Systems (BBSs). They originally started out as electronic versions of the old community bulletin boards you might have seen at the grocery store or co-op. Originally, BBSs were limited to exchanging electronic messages between people within a community. Today, some can connect you to the Internet.

A typical BBS may consist of nothing more than a personal computer and one or two modems and phone lines. BBSs are frequently operated by individuals right in their own homes. Some charge a fee, but it is usually small; many are free.

A computer, a modem, and a phone line are about all you need to set up a BBS, so anyone can do it. Since they are easy to start, there is little consistency in quality and services offered. Some BBSs are so well run they're hard to distinguish from the "real" online services. (Steve Case, president and founder of America Online, supposedly got his start with a BBS.) Others are poorly run and a waste of (your) time.

If there's a BBS in your area, chances are the best way to find it is through word of mouth. Check with local colleges, schools, electronics stores, extension agents, and other computer users in your community. You can also use online services and the Internet itself to help you in your search. On AOL, you can check out their BBS Corner for general information about BBSs, and you can search their database for a BBS by state or area code. On CompuServe, type GO ASPFORUM for a list of BBSs.

INTERNET CONNECTIVITY OVER A BBS

Most BBSs don't offer any Internet services. Instead, they have their own networks (like FidoNet) that they use to exchange e-mail and other files with other BBSs around the country. Some BBSs, however, do provide some Internet services, but there is quite a bit of variation in which services they offer.

It used to be unusual for a BBS to have a full-time "live" connection to the Internet. Instead, like many FreeNets, most BBSs connected to the Internet once a day to pick up and drop off e-mail from the BBS's subscribers. The few BBSs that were "live" on the Net offered only shell accounts, which meant no World Wide Web. But today, more and more BBSs are now offering full Internet connectivity, including Web access.If cost is a critical issue for you, you may find that a BBS with a free local number and some kind of Internet access may be a better deal than a full-service ISP that you have to call long distance to reach. A BBS is usually more community-oriented, and many people prefer the warmth of a locally operated BBS (or FreeNet) to the coldness of the huge, global Internet. By providing more than just Internet access, BBSs (and especially FreeNets) help knit communities together—electronically speaking.

BBS SOFTWARE

You probably won't need to buy any software to access a BBS. The basic text-based communications program that came with your modem (such as ProComm, Microphone, Hayes Smartcom, or Microsoft Terminal, and so on) will let you log onto most BBSs. Some of the fancier BBSs may offer special graphics software that makes their BBS look more like Windows or Macintosh. If this is the case, they'll probably have the software you need (and they'll usually let you download it free, too).

When you first log onto a BBS, you may be asked a lot of confusing questions about configuring your system, such as whether you want ANSI graphics. If you don't know the answers to these questions, then accept the BBS's default answers, and everything should work. Once you get online, you can use their Help system to get information, then try different settings to make things work more smoothly.

If you are trying to log onto a new BBS and you don't know the communications parameters they use, try 8 data bits, no parity, and 1 stop bit (8-N-1). This combination works most of the time.

OPERATING PRACTICES THAT CAN SAVE MONEY

After you've chosen the least expensive method you can find for connecting to an ISP, the next step is to learn operating practices that will reduce the number of hours you'll need to spend online to get things done.

If you are paying $19.95 a month, have free local access, and are getting unlimited hours of "free" connect time each month, you don't need to watch the clock. Many Internet users (in cities) can afford to spend several hours per day surfing the Net, investigating "interesting" Web sites, reading lots of newsgroups, chatting in online forums, and downloading lots of free software they'll probably never use. It's not an efficient use of their time, but it's a cheap (and relatively harmless) way to spend an evening. But such profligate usage of the Internet has contributed to the myth that you actually need to spend 20 or even 40 hours a month connected to the Internet to really take advantage of this new communications resource. This is simply not true. You can certainly take full advantage of the Internet's bounty in around five hours per month (or less) if you have a high-speed (28,800 bps or faster) connection and use your time efficiently.

WORKING ONLINE VERSUS OFFLINE

The key to using the Internet efficiently is to be online (with the clock running) only when you are actually sending and receiving information. Note that this is not how most urban users use the Internet. Since urban users are paying only a few cents (or even a fraction of a cent) per minute, they can afford to stay online while they leisurely read their e-mail and compose their replies. They can stay online while they print and even while they go out for a cappuccino. But if you are paying long-distance (or toll-free) rates to connect, you can't afford to spend a great deal of time online. But you don't need to do that to make effective use of the Internet.

It is necessary to be online to receive e-mail, but you don't need to be online to read it. Likewise, you must be online to send e-mail, but you don't need to be online when you write it. Most things you'll want to do regularly on the Internet can be done quickly. Therefore they can be done cheaply, if you go offline to read, write, and print. This is very easy to do: once you've downloaded the e-mail (or news group) messages or have loaded several Web pages, disconnect (log off) from the Internet and read the messages or Web pages at your leisure.

If you work offline, then how much time do you really need to spend connected to the Internet? The answer to that question depends on what you want (or need) to do. We can't answer that for you, but we can create a hypothetical list of typical Internet activities and show you how these tasks can be performed more efficiently and cheaply by going offline.

TRY THIS SIMPLE EXPERIMENT

If you have an external modem, go online and watch the status lights. You'll see for yourself which tasks you have to be online for and which you don't. When you are sending data over the phone, one of the modem lights will flash. (This light is usually labeled SD or TX.) When you are receiving data over the phone, a different light will flash (usually labeled RD or RX). Note that some lights will always be on, and some will always be off—but when no light is flashing, you are not transmitting or receiving any data. Sometimes the status lights are not flashing because you are waiting for the other computer to respond. In that case, you'll have to remain online until the other computer responds. (This shouldn't take more than a few seconds.)

So, look at the modem's lights the next time you go online and retrieve your e-mail. During the initial logon phase, the lights flash as your computer and your ISP exchange information like your name and password. Then the RD light will flash as your e-mail messages are downloaded to your computer. At this point, the messages are stored in your computer. As you read the e-mail, the modem's lights aren't flashing. (An occasional flash may mean the system is checking to make sure you're still there.) When the lights aren't flashing, you're not getting or sending any data, although you are online and paying for that privilege by the minute. Next, print a message on your printer. Notice that the modem's lights don't flash. Save the message to disk. The modem's lights still don't flash. You can read the messages and type a response, and the modem's lights don't flash then either. Only when you "send" the e-mail responses back to the Internet will the SD light flash, signifying the fact that data is flowing through the modem.

You had to be online to initially receive the e-mail messages. You had to be online to send the replies. But you didn't have to be online to read, print, or save those messages. You didn't have to be online to write the response either.

It can easily take 5 minutes (or more) of online time to retrieve a message, read it, and then write a response and send it. On the other hand, it probably takes less than a minute of online time to actually get the e-mail message and perhaps another minute of online time to send the response. The 3 minutes (or more) spent reading and typing can be done offline at no cost.

If you go online and retrieve your e-mail and then log off, you can read, print, save, and type the response at no cost. After you finish writing your response, you can go back online and send it, paying to be online only when you really need to be.

To see the effect of working offline versus online, we've created a hypothetical set of activities that a heavy Internet user (with lots of free time) might do every business day. Then we calculated the approximate amount of time each activity requires. Obviously, e-mail messages and other files come in various lengths, and people differ greatly in how fast they read and type. But we believe these times are fairly typical and, in any case, they illustrate the point.

Activity	Time Spent Online (minutes)	Time Spent Online/Offline (minutes)
Read and respond to e-mail	5	2
Read a newsgroup	10	1
Read an online magazine	15	2
Check the weather	5	1
Download a few Home Pages	10	2-4
Total Daily Activity	45 min.	8 to 10 min.
Total for the month	15 hours	2.5 to 3 hours
Cost	$100	$25 to $30

(Assumes daily use for 20 business days @$9.95 for 5 hours of Internet service, plus a long-distance rate of 10 cents per minute.)

These calculations assume that both users have a 28,800 bps connection. And we're also assuming consistently heavy use, but look at the price difference! Obviously, the exact amount of time that these activities take will depend on the length of the messages and other files, how fast you read, how fast you type and, of course, any distractions. But you can easily reduce your monthly Internet costs by more than two-thirds to three-fourths by being online only when you really have to be.

UNAVOIDABLE OVERHEAD

Depending on the access provider, you can expect to take at least 30 seconds to a minute to log onto the Internet, and during this time you are paying a long distance charge (even if you're not fully connected to the ISP yet). Some online services may be a little slower with the logon procedure than the full-service ISPs, but make up for it with FlashSessions and other features that speed you up when you're online. In any case, it will take roughly 30 seconds or so to connect. This initial overhead is for the most part unavoidable and is due to the electronic handshaking between the modems and the ISP's logon procedures. Since it's going to take about 30 seconds to access the Internet each time you log on, you should plan ahead and make the most of each session. The rule of thumb is to log off if what you are doing will take substantially longer than 30 seconds (for example, reading several e-mail messages). Otherwise, you should stay online until the job is done.

WAITING FOR BAUDOT

You've got a blazing new 220 MHz Pentium processor and a lightning-fast 28,800 bps connection. You're all set to race down the Information Superhighway when, suddenly, you see the yellow flag. Traffic on the Information Superhighway slows to a crawl, for the same reason it slows on the real superhighway—too much traffic or too many collisions.

Computers, like cars, can crash. Data can take a wrong turn and get lost. Traffic conflicts can jam an intersection. In short, just about anything that can happen on the real superhighway can happen in cyberspace (except of course, no one really dies on the Information Superhighway).

Since you're paying by the minute, these slowdowns increase your costs. Knowing how to detour around these virtual traffic jams can save aggravation, time, and money. But to find a solution, you've got to know what—or more precisely, who—is causing the problem.

OTHER OPERATIONAL STEPS YOU CAN TAKE TO REDUCE YOUR COSTS

E-MAIL

To use e-mail efficiently, read and write it offline. You can do this by writing your messages (using an e-mail program or word processor) before you go online. When you do go online to send a message, check to see whether you have received any e-mail messages, then log off and read them. If you need to reply to any e-mail you've received, you can write the reply, log on again and send it. Or, just wait until tomorrow and send the reply when you are checking for new e-mail.

Some e-mail programs have a feature called Check on Send, usually found in a setup or preferences file. With Check on Send, the e-mail software automatically checks for new e-mail every time you send any. It only takes a second or two more to check when you are online sending e-mail, which is less time than it takes to log on and check to see if you have any e-mail.

NEWSGROUPS

Newsgroups are notorious time wasters. Be choosy about which ones you subscribe to. Generally speaking, it may take more time to read your newsgroups offline. Newsgroups can get large numbers of postings, and most people usually don't read them all. It may take longer to download the entire file than to stay online and read the few you want to see. The answer depends on the size of the file and how many you really read. When you initially subscribe to a newsgroup, every one of the previous messages will be marked unread. So if you download the entire newsgroup file, you'll get every posting, which could take quite awhile. Instead, when you first subscribe to a newsgroup, read a few while you are online, then mark them all as read. The next time you check your newsgroup, you'll only see the new unread messages which you can download and read offline. Some newsreaders have a read offline function that lets you specify (usually by keyword) which postings you want or don't want to download. With this feature, you could subscribe to a fairly general newsgroup related to veterinary medicine and download only the messages you want to be read offline.

MAILING LISTS

The best thing about mailing lists is that the information is automatically sent to you (after you subscribe). Mailing lists are usually packets of separate e-mail messages, so they're easy to retrieve and read offline, if there aren't too many of them. Be choosy about which lists you subscribe to and don't hesitate to unsubscribe if all you're getting is junk e-mail.

THE WEB

You can spend a lot of time and money surfing the Web. But you can also find information on the Web that is truly worth the expense. Here are some suggestions that will help you get what you need from the Web without spending too much.

GRAPHICS ARE OPTIONAL

Because the files are so much larger, graphics take longer to download than text. A 1-inch-by-1-inch picture might be 10 kilobytes or bigger, while the entire text of a document might be only 1 kilobyte. Most Web browsers will let you turn off the graphics. In Netscape, for example, look under the **Options** pull-down menu and uncheck **Auto Load Images**. In Explorer, pull down **View**, **Options** and select **General**. Now, uncheck the **Show Pictures** box. Now when you click from page to page you'll skip the graphics and only download the text. At each place in the text where a graphic would be, you will see the little *graphics icon* illustrated on the left.

If you are surfing the Web with "graphics off" you can always click on the graphics icon and download that particular picture or image.

If you want to see that particular image, you can click on its graphics icon, and it will be downloaded. Sometimes graphics add to the information. At other times, they are cute, but useless. Occasionally, they are just useless. Since you are paying to download them, and they are optional, turn off the graphics if you don't think they are adding value to the material.

Online services, particularly AOL, use a lot of graphics and you can't easily turn them off.

Sometimes a whole page consists of nothing but a large graphic, in which case you'll have to download it to see what it says. Usually these are image maps that you can click on to go somewhere else. They can sometimes be faster than sloshing through a lot of menus. In some cases, pages that use a lot of graphics have text-only versions available. So instead of downloading the graphic image, just click on the text-only link.

HOME PAGE

Most Web browsers come with a default home page, usually the home page of the company that wrote the browser software. With millions of copies of software out there, a fairly large number of people are logging onto those servers at any one time to get the default home page. It doesn't take that much time to load a home page, but over the course of a year, those few wasted tens of seconds each day can total up to hours. The smart thing to do is to change your default home page to one that is more useful (such as *Farm Journal Today* at **www.farmjournal.com**).

JANE! STOP THIS CRAZY THING!

Netscape and other Web browsers have a little stop sign (or similar icon) that you can click anytime you want to stop a download. Online services have similar cancel commands. These change from time to time and can depend on the type of computer system and software you are using. Check with your online service for more information.

BOOKMARKS AND FAVORITES

Most Web browsers will let you store the URLs of various Web sites you visit so you don't have to search for them or type them in again to make a return trip. Netscape calls them Bookmarks while Internet Explorer calls them Favorites. By saving the addresses of sites you may want to visit again, you save much time and you don't risk any typos.

PRINTING

Because of the computer overhead that printing requires, it's a good practice to wait until you are offline before printing. Most Web browsers automatically store e-mail messages on your computer and have a special caching feature that captures Web pages and stores them, so you can print an e-mail message or a Web page after you log off.

COMPRESSION

Compression is another time saver. Some modems have the ability to compress data "on the fly." Compression by modem can (in theory) achieve impressive throughputs if telephone line conditions allow. Modems with built-in data compression can achieve very high data transfer rates (57,600 bps and higher) at raw connection speeds of 28,800 bps. This on-the-fly compression works only if both you and the ISP have it and if the file is "compressible."

Another way to use compression is to pre-compress software files before they are transmitted by modem. Once you've downloaded the file, you then use a decompression program to expand a compressed file back to its original size.

Compression and decompression programs come in pairs. For example, if the file was compressed with StuffIt, you'll need StuffIt Expander to decompress it. Matching decompression programs are usually available online in the same place you found the compressed file. Most online services provide customers with the more commonly used decompression programs (and some even automatically decompress any files you downloaded as soon as you log off). A few compressed files are "self extracting" and do not require a separate program to expand them. The names of self-extracting files usually end with the letters "exe" or "sea." The first time you attempt to run them, they self extract, which means that they decompress themselves.

OPERATING TIPS FOR ONLINE SERVICES

Because the major online services like America Online and CompuServe serve as sort of friendly intermediaries between the user and the cold, hard world of the Internet, you'll probably experience few operational troubles if you use a major online service as your Internet provider. All of the major online services provide their own software, which is easy to install and generally trouble free. They also update their software on a regular basis, so you can always operate with the latest version. For these and other reasons, it's rare to hear of an operational problem with one of the major online services. There are many full-service ISPs of that caliber, but some ISPs are little more than two guys in a garage with a couple of PCs. If their garage happens to be a local call, then you're probably stuck with them for economic reasons. If you're calling long distance, then having trouble making a connection or a problem with flaky software is more than just an aggravation, it's also an unneeded expense.

OFFICIAL GUIDES

Most major online services have an "official guide" or other handbook that explains in detail how to use the online service's features and how to use them efficiently. Each online service has its own peculiarities that could stump you, as well as time-saving features that also save you money, if you know how to use them. Since features vary from service to service, it's a good practice to get the official guide for the online service you'll be using. Most of the services promote their guide books online from time to time, so you're bound to run into an ad sooner or later. You may also be able to buy them from the service's "online store."

Instead of buying the official guide, you could use the online service's help system. Most online services don't charge you when you're getting help, but this policy applies only to the online charges; toll-free surcharges and long-distance charges, if any, still apply. So it may be cheaper in the long run to buy the service's "official" online guide.

GRAPHICS

The major online services tend to use a lot of pictures and graphics, because clicking on a picture is easier than reading a lot of text, and having lots of graphics helps make their service look good. As a rule, turning off the graphics from most online services is not an option, except when you are using the Web browser. However, most online services are generally efficient with their use of graphics and there are other, more effective ways to bypass all those screens.

With many online services, some of the pictures and graphics you need come on the installation disk. When you first sign on, the first few minutes are spent updating icons, pictures and graphics that were created after the disk was made. It can sometimes take quite awhile to get all the new graphics, but keep in mind that this is a one-time event. The next time you sign on, you'll already have most of the pictures and icons on your hard drive, so you won't have too many new ones to download. (But there will always be some.)

KEYWORDS AND OTHER SHORT CUTS

Some online services let you use short cuts to go straight to an area of interest without clicking through stacks of menus (and waiting for the graphics to download). AOL calls these shortcuts keywords; CompuServe calls them GO commands.

Since menus are useful when you're learning how the system works, you'll probably need to use the menus at first. But when you find an area you plan to

▼ *Appendix B*

return to, write down its keyword or Go command. (It's usually at the bottom of the active window). Using AOL, for example, you would click the **Keyword** button, type in the keyword and you're there! Most of the major services have similar time-saving features. Get a copy of your online service's guide and learn how to use these time (and money) saving features.

IN THE STILL OF THE NIGHT

The time of day you access the Internet can also have a big impact on costs. Obviously, if you are using long distance, calling in the evening or on weekends when rates are lowest will result in substantial savings. Most long-distance plans cut their rates in half after 6 p.m. and on weekends. However, many ISPs and online services report that their busiest hours are at about the same time—early evening and on weekends. Coincidence?

Although your long-distance cost may have dropped, slow performance by an ISP or online service can rob you of some of that savings. It is sometimes better to sign onto the Internet later in the evening (say after 8 or 9 p.m.) rather than right at 6:01 p.m. when all those other people who just got off from work are signing onto the Internet. Early morning hours are even better (assuming you sign on before long-distance rates go back up–typically at 8 a.m.) because few people get up early to use the Internet.

Internet servers are also likely to be at their fastest late in the evening or early in the morning. There are exceptions, of course; the moment you sign on might be just the time the company has decided to run its monthly payroll. Still, if you plan to do some intensive searching and retrieving, wait until Letterman is on (or until early morning) to get faster responses and lower costs.

APPENDIX C
PCS AND MACS FOR THE INTERNET

If you are in the market for a new PC or Mac to access the Internet, there are a few things you should know. First of all, buying a PC is not like buying a tractor. My grandfather had a Ford 8N, circa 1952. A few years ago, I saw a well-worn 8N on a used car dealer's lot and bought it (for nostalgia reasons). New Holland still carried parts and the original repair manuals. The 8N didn't have live hydraulics and needed an add-on PTO clutch for safety, but it had a three-point hitch, and if power needs weren't critical, it worked fine (after a major rebuild). After forty-odd years, you could still hook up the latest attachments and that old 8N would run them.

Today, if you asked for advice on buying a small tractor to pull, say, a five-foot flail mower, we could suggest a tractor of between 30 and 40 horse power, and you could easily find one. You could get an old 8N and fix it up, or you could get one of the new Boomer's from New Holland. Of course, the Boomer has an efficient diesel engine, live hydraulics, ROPS, adjustable seats and steering and a host of new features, but at 34 horsepower this little tractor has a similar performance envelope to the old Ford 8N series. You'd be happier (and safer) with a new tractor like the Boomer, but either will let you attach and pull a five-foot flail mower. So if the only requirement is that the tractor have around 30 to 40 horsepower, then in that sense, either will "work." In any case, your New Holland, Case IH, or John Deere dealer would have some model tractor in that horsepower range. They've got one today, and they probably will ten years from now.

With computers, it's another story. We can't recommend the precise "horsepower" for a computer for the Internet because by the time you are reading this, they won't be making that model of computer anymore. If we said that a PC with a 100 MHz Pentium processor would be a good "starter system," by the time you called to order, a 166 MHz Pentium MMX processor would be the slowest available. Six months later, a 200 MHz might be the slowest you could find.

Fortunately, just about any new computer sold to the public today (and certainly tomorrow) will meet and, in virtually every case, exceed all the requirements for accessing the Internet. As the technology continues to advance, what was a "middle of the road" system six months ago is now considered "low end" or is out of production altogether. That doesn't mean these systems weren't good, it just means that computer technology is advancing so rapidly. Below, we've listed the minimum specs for systems that should meet all the requirements for your Internet applications. By the time you contact your computer dealer, you'll probably be able to buy a faster computer, with a bigger hard drive, and more memory for less money!

WINDOWS-BASED PC

Processor: 166 MHz Pentium w/MMX
Hard Disk: 1.6 to 3.2 GB Hard Drive
Memory: 32 MB RAM
CD-ROM: 8X Speed (or faster)
Video Card: 4 MB Video RAM
Monitor: 15-17 in.
Sound Card: Optional, preferably 16 bit (or bigger) and Sound Blaster compatible.
Modem: 33.6 kbs (U.S. Robotics makes a nice one).
Software: Microsoft Windows 95 (or 98) & MS Office '97 (Small Business Edition)
System Price: about $ 2000

Printer: Optional (HP makes a nice color inkjet printer that sells for less than $300).

PC VENDORS:

Dell Corporation
Contact Info: 1-800-WWW-DELL or http://www.dell.com
Gateway Computer
Contact Info: 1-800-846-4208 or http://www.gw2k.com
Sony Computer
Contact Info: 1-800-352-7669 or http://www.sony.com
MicroWarehouse
Contact Info: 1-800-696-1727 or http://www.warehouse.com/
Best Buy
Contact Info: 1-800-369-5050 or http://www.bestbuy.com
Circuit City
Contact Info: http://www.circuitcity.com

MAC OS

Processor: 180 MHz 603e or 604e PowerPC Processor
Hard Disk: 1.6 to 2.4 GB
Memory: 24 MB RAM
CD-ROM: 8X Speed (or faster)
Video: 2 MB Video RAM
Monitor: 15-17 in.
Modem: 33.6 kbs (U.S. Robotics makes a nice one).
Software: Mac OS 7.6 or OS 8
System Price: about $ 2000

Printer: Optional (HP makes a nice color inkjet printer that sells for less than $300).

MAC VENDORS

Mac Warehouse
Contact Info: 1-800-397-8508 or http://www.warehouse.com/MacWarehouse/
UMAX Computer Corporation
Contact Info: 1-510-226-6886 or http://www.supermac.com
Apple Computer Corporation
Contact Info: 1-408-996-1010 or http://www.apple.com
Power Computing
Contact Info: 1-800-370-7693 or http://www.powercc.com

USED PC'S AND MACS

If you are buying a used computer to access the Internet, here are our suggested "minimum" requirements. Keep in mind that you can cruise the Information Superhighway with less, just as you can drive on the Interstate in a Yugo. If you do buy a used PC or Mac, make sure you get a new modem (at least a 28.8 kbs, although a 33.6 kbs would be better).

IF YOU'RE USING WINDOWS 95, LOOK FOR ...

Processor: 66 MHz 486 or equivalent
Memory: 16 Mb RAM
Hard Disk: Minimum 520 MB
CD-ROM: 4X speed (or faster)
Video Card: 1 MB Video RAM
Monitor: 15 in.
Sound Card: Optional, preferably 16 bit (or bigger) and Sound Blaster compatible.
Modem: 28.8 kbs

IF YOU'RE USING WINDOWS 3.11, LOOK FOR ...

Processor: 25 MHz 486 or equivalent
Memory: 8-16 Mb RAM
Hard Disk: Minimum 250 MB
CD-ROM: 4X speed (or faster)
Video Card: 1 MB Video RAM
Monitor: 15 in.
Sound Card: Optional, preferably 16 bit (or bigger) and Sound Blaster compatible.
Modem: 28.8 kbs

IF YOU'RE USING MAC OS 7.5, LOOK FOR ...

Processor: 100 MHz PowerPC 601 Processor
Memory: 16 MB RAM
Hard Disk: 520 MB
CD-ROM: 4X speed (or faster)
Video: 1 MB Video RAM
Monitor: 15 in.
Modem: 28.8 kbs

APPENDIX D

USING INTERNET SOFTWARE WITH WINDOWS 3.1

First, you'll need to obtain Internet communications software. This is not the browser such as Netscape Navigator, but separate programs that Netscape uses to have your computer dial your Internet Service Provider and communicate over the Internet. There are two pieces of software you'll need: one is called the TCP/IP stack and the other is called the dialer. Many ISP's provide this software free of charge, and we would strongly recommend that you check with them first to see if they will supply this software. However if your ISP doesn't, there are several options available.

A few years ago, browsers didn't come with a TCP/IP stack and a dialer built in and users had to obtain and configure stand-alone programs such as Trumpet Winsock. Getting different programs from different manufacturers to work together harmoniously was sometimes a problem. Today, both Netscape and Internet Explorer make special versions of their browsers that have the TCP/IP stack and dialer built in. At this writing, the Netscape version with everything built in is Netscape Navigator 3.01 w/ Modem software. The Internet Explorer version is Microsoft Internet Explorer 3.02a with TCP/IP stack and dialer. These product numbers may change, so check with your computer software dealer and make sure the version of Netscape or Explorer you get is designed for Windows 3.1 and includes the stack and dialer (sometimes called the modem software).

While both Netscape Navigator and Internet Explorer can hook you up with an ISP, it's probably better for people who live in rural areas to find an ISP and set up their own account (see Appendix B). If you set up your own ISP account, you'll need to have some information from your ISP in order to properly configure your Internet software. There's a form at the end of this section to record the information you'll need to know.

OBTAINING NETSCAPE NAVIGATOR 3.01 W/ MODEM SOFTWARE

You can buy Netscape Navigator 3.01 w/ Modem software from most computer software dealers or mail order companies. If you can use a friend's computer to access the Net, you can download it from Netscape's Web site. Their Web address is **http://home.netscape.com/comprod/mirror/client_download.html**.

Once you're on their site, select "Navigator 3.0 plus Modem Software - Floppy Install" as the desired product. Select Windows 3.1 as your operating system, and then select your location and preferred language. Next, download and save the file from any of Netscape's many mirror sites. After the file is downloaded, move the file into an empty directory and run it. The file will create seven directories labeled disk1, disk2, disk3... disk7. (Thus, you will need about seven floppy disks to hold the software so you can take it home and install it on your computer.) Copy the contents of the directory 'disk1' to the first floppy, the contents of 'disk2' to the second floppy and so on. Now take the seven floppies you've just created home, and run the setup program (on disk 1) on your own computer and it will install the software.

CONFIGURING NETSCAPE NAVIGATOR

Once the software has been installed, open the Netscape Dial-Up Program Group and double-click the Account Setup Wizard icon. Netscape will now ask you if you have an existing Internet account, or if you need to sign up with an ISP for a new account. Although Netscape can set up an Internet account with an ISP, it's probably a good idea to have already done this yourself.

As part of the installation process, Netscape will want to know the name of your Internet Service Provider and your name. If your Internet Service Provider supports PAP—the password authentication protocol—you may enter your account username and password now, and Netscape will remember them from now on. If your ISP doesn't support PAP, just check the "I will need a Login Window" to open a terminal window each time you dial. Netscape also needs to know the number your computer must dial to connect to the ISP (hopefully, this will be a local number). If you have call waiting, you should disable it while you're online. Simply check the "Disable Call Waiting" box, and enter the code your phone company requires to disable call waiting. Usually, the code is *70, but check the front of your phone book to be sure.

Netscape will also need to know, your Internet Service Provider's IP address (for their DNS server). This address is a series of numbers separated by periods as in 128.163.192.9. Your ISP may also have an alternate IP address, but this secondary DNS address is usually not required.

Finally, enter your Internet Service Provider's POP, SMPT and NNTP servers' names. These tell Netscape where your incoming and out going e-mail messages are to be stored and where your newsgroups are located. If you don't know this information, you can enter it later using Netscape's Mail and News Preferences from the main Netscape browser window. After you're done,

Netscape will open the Modem Setup Wizard to determine which brand of modem you have and to which port it is attached. After selecting a modem, you are ready to go. Just double-click the Netscape icon to go online!

To have Netscape save your e-mail password, click Options, Mail and News Preferences, then click the Organization tab. Then, click the checkbox next to "Remember Mail Password" and click OK.

OBTAINING INTERNET EXPLORER 3.02A PLUS DIALER SOFTWARE

Like Netscape, Internet Explorer 3.02a plus dialer software is available from any software retailer in the form of the Internet Explorer Start-Up Kit. Unlike Netscape this software is not easy to get via the Web, because the installer file is too large to fit on a single floppy and (as of this writing) it doesn't automatically break itself into separate files that will fit on several floppy disks. You can get around this problem if both you and your online friend have large 'Zip' drives or you are adept at creating "disk-spanning" archives with WinZip (available at **http://www.winzip.com**). In that case, download the software on your friend's computer by going to Internet Explorer's download site at **http://www.microsoft.com/ie/download**. Now, select your operating system (Windows 3.1) and click **Next**. Now, select Internet Explorer 3.02a for Windows 3.1 and click **Next**. Now, select the language—Internet Explorer 3.02a (Browser, Stack/Dialer, Mail, & Java) for Windows 3.1 and click **Next**. Finally, download the file to your online friend's computer from one of Microsoft's mirror sites, then take the executable file to your home computer and install.

CONFIGURING INTERNET EXPLORER'S DIAL-UP SOFTWARE

After you've installed the program on your computer, simply double-click on the **Get on the Internet** icon in the "Microsoft Internet Explorer" Program Group. This brings up Microsoft's Internet Connection Wizard which asks you if you want to manually specify your Internet Service Provider settings or use Microsoft's Automatic Internet Service Provider Referral System. If you already have an account with an ISP, select **Manual**. If you want to use Microsoft's referral service to set up an account for you, select **Automatic**.

USING MICROSOFT'S AUTOMATIC INTERNET REFERRAL SERVICE

After selecting **Automatic**, Microsoft's dialer will ask you for your area code and the first three digits of your telephone number. Based on that information, Microsoft will download a list of Internet providers that are local to you (if any). If you live in a rural area, there may be none on the list, so it's often better to find a local ISP on your own and to enter the information manually.

MANUALLY ENTERING ISP INFORMATION

By checking **Manual**, Microsoft's Internet Connection Wizard assumes that you already have an account set up with an ISP. The Connection Wizard will ask you for the ISP's name, telephone number, and your username and password. Next, Internet Explorer will want to know whether you have a static or dynamic IP address. If you're not sure, just tell Internet Explorer that "My Internet Service Provider Automatically Assigns me one" and continue. Now, enter your ISP's primary DNS server IP address in the space provided. This address is a series of numbers separated by periods as in 128.163.192.9. Next, the Wizard asks you if you'd like to set up your Internet mail account now. If you select yes, you will be asked for your mail username and password (usually it's the same as your login ID and password) as well as your full name, e-mail address, and your ISP's POP and SMTP server. Next, you'll be asked to set up your newsgroup account. To do this, enter your name, e-mail address, and your newsgroup's server name. (If your newsgroup requires a username and password, you may enter those too.) With all that information entered, you're ready to go. Just click on the **Internet Explorer** icon to connect.

HERE'S A CHECKLIST FOR THE INFORMATION YOU'LL NEED TO GET FROM YOUR ISP:

ISP's Domain Name:_____

ISP's Local Dial-in Number:_____

ISP's Domain Name Server (DNS):_____

ISP's News (NNTP) Server:_____

ISP's POP Mail Server:_____

ISP's SMTP Server:_____

Your user ID or username:_____

Your password:_____

Your e-mail address:_____

ISP's home page:_____

ISP's telephone number
(voice line for tech support):_____

ISP's e-mail address:_____

APPENDIX E
USING INTERNET SOFTWARE WITH WINDOWS 95

First, check to see if Dial-Up Networking is installed on your computer. This program comes with Windows 95, but is not normally installed automatically. You check to see if Dial-Up Networking is already on your computer by double-clicking on the **My Computer** icon. If the window that opens up contains a folder called "Dial-Up Networking," then the software has already been installed. If you do not see that folder, then you must install Dial-Up Networking. To do this, click **Start**, then go to **Settings**. Now, click on **Control Panel**. Once the **Control Panel** opens, double-click on the icon labeled **Add / Remove Programs**. After the **Add / Remove Programs** window opens, click on the tab at the top labeled **Windows Setup**. Then, double-click on the icon labeled **Communications**. In the **Communications** window, check the boxes beside "Dial-Up Networking" and "Phone Dialer" and click **OK**. Finally, click **OK** in the **Add/Remove Programs** box. Windows 95 will now Install Dial-Up Networking software on your computer. (Note: You may be prompted for your Windows 95 CD-ROM or Setup Diskettes.) After the software has been installed, reboot your computer.

Once Dial-Up Networking has been installed, you should check to see if TCP/IP has been installed. To do this, click **Start**, go to **Settings**, then **Control Panel**. Inside the **Control Panel**, double-click on the icon labeled **Network**. After the **Network** window opens, click the **Configuration** tab and you will see a list of all the networking components installed on your system. Look through the list for an icon labeled TCP/IP Protocol. If you see it, TCP/IP is already installed. If you do not see a TCP/IP icon, you must manually install it. This is done by clicking on the **Add** button. Once you do this, Windows will open a window called Select Network Component Type. Double-click on the icon labeled **Protocol**. In the **Protocol** window, you will see a list of manufacturers and a list of protocols. Click on the icon labeled **Microsoft**, then select the **TCP/IP Protocol** and click **OK**. Finally, click **OK** in the **Network** window. Windows will then install the files needed for the TCP/IP Protocol. (Note: You may be prompted for your Windows 95 CD-ROM or diskettes.) When prompted, choose to reboot your computer.

SETTING UP AN INTERNET CONNECTION

Once Dial-Up Networking and TCP/IP Protocol are installed, you're ready to set up an Internet connection. At this point, you should already have an Internet account with an ISP (see Appendices A and B). The following steps require you to enter information about your ISP and your Internet account. You can use the form at the end of this section to record the information you'll need to complete these steps.

To set up your Internet connection, double-click the **My Computer** icon. Then in the **My Computer** window, double-click on the icon labeled **Dial-Up Networking**. Now, double-click the **Make New Connection** icon. In the window that pops up, type in the name for the Internet Service Provider with whom you have an account. Then, select a Modem Type from the pull-down menu. (Note: If you don't have a modem installed, Windows 95 will prompt you to install one using the Add Modem Wizard.) Once you've selected a modem, click **Next**. You will then be prompted for the telephone number of your Internet Service Provider. Enter it, then click **Next**. To save this information, click **Finish**. Now, go to the Dial-Up Networking folder. There you will see a new icon with the name of the ISP you just entered. Right-click on this new icon, and select **Properties**. In the window that pops up, you'll see the telephone number you just entered, as well as the modem you selected. Now, click on the button labeled **Server type**. Make sure the Type of Dial-Up Server specified in the pull down menu is "PPP: Windows 95, Windows NT 3.5, Internet," and make sure the box next to "TCP/IP" is checked. Now, click the button labeled **TCP/IP Settings** and select the **Server Assigned IP Address** button. Now, click the **Specify Name Server Address** radio button. In the field labeled "Primary DNS," enter the IP address of your Internet Service Provider's Domain Name Server. This address is a series of numbers separated by periods as in 128.163.192.9. (If this address isn't right, nothing will work.) Finally, click **OK** on all the open windows to save your information.

Now that you've told the computer all about your ISP, you'll need to tell it some things about yourself. It needs to know what type of password authentication system your ISP uses. If they use PAP, just fill in your username and password in the blanks Windows 95 provides when you dial. If they don't use PAP, or if they don't know, right-click on the connection as above, then select **Configure**. In the window that pops up, select the **Options** tab. Then check the box next to "Bring up Terminal Window After Dialing" and click **OK**.

CONFIGURING THE BROWSER

In addition to the above mentioned communications software, you'll also need a program called a browser. Your system may have come with a version of Internet Explorer (Look for a globe and magnifying glass icon on your desktop. Usually this icon is labeled "The Internet" but may also be called "Internet Explorer.") or Netscape. (Look in your **Start** menu under **Programs**, for a folder called **Netscape**.)

If you can't find either of these browsers, you'll need to get one. There are a couple of ways to go about this. First, visit any software/computer store and purchase a copy of Netscape Navigator/Communicator or a copy of the Internet Explorer Starter Kit (about $24.99). Either of these two kits will contain everything you need to surf the Web from Windows 95 (or Windows 3.11). Second, you could visit a friend who is already connected to the Internet and ask them to download a browser for you. If you follow this route, you must find a way to move the file from their computer to yours. This can be tricky, since most browsers will not fit onto a single floppy disk. It can be accomplished, however, via a Zip Drive, or with a floppy disk spanning archive (created with WinZip or PKZip). If you want to try to download a browser from the Web, use the instructions in Appendix D but change the operating system type to Windows 95. Also, check with your intended ISP to see if they have a browser they can send you.

CONFIGURING YOUR BROWSER

It seems that every few weeks or so either Netscape or Microsoft is releasing an "improved" version of its browser. While some of the "improvements" may be real and even necessary, it also means that some of the buttons get moved around and the names of some of the buttons change. This makes it hard to give you step-by-step instructions on how to configure a browser, because we can't know where Netscape and Microsoft are going to put the buttons and what they are going to call them in the future. The following are generalized instructions that should work in most cases.

CONFIGURING NETSCAPE

To use Netscape Navigator 3.01, you shouldn't have to change any of the default settings. Right out of the box, the software is (or should be) properly configured for accessing the Web. All you need to do is customize it to suit your tastes. For example, to save space on computers with low resolution monitors, you can tell Netscape to display its toolbar as pictures only, instead of pictures and text.

To use Netscape for e-mail, you will need to enter some information about yourself and your e-mail account. To begin, click **Options** then **Mail and News Preferences**. In the window that opens, click the tab labeled **Servers** and enter your SMTP and POP3 server names, as well as your POP user name in the blanks provided. Your POP user name is simply your e-mail user name. The POP3 and SMTP server information must be provided by your ISP.

To access news groups, you'll need to set up the News portion of the browser. In the section at the bottom labeled "News," type in the name of your Internet Service Provider's News server. (Although this information isn't required to send and receive e-mail, it is required to use Netscape to read and post to newsgroups.) This News server name must also be provided by your Internet Service Provider. Now click the tab labeled **Identity** and enter your full name and your e-mail address in the blanks provided for you. (You may also enter an organization name, or a return address, if you want the replies sent to a different e-mail address.) Now, click **OK**. You're ready to use Netscape to send and receive e-mail and to read newsgroups.

To activate the e-mail system, just click on the little envelope symbol at the bottom of the Netscape window. Netscape will then bring up an e-mail window and ask you for your e-mail password.

CONFIGURING NETSCAPE COMMUNICATOR 4.0

Essentially the same as Netscape Navigator 3.01, except you click **Edit**, then **Preferences** to view the **Mail and News Preferences**. Now, enter the information provided by your ISP.

To have Netscape remember your e-mail password, click Options, Mail and News Preferences, then click the Organization tab. Click the checkbox next to "Remember Mail Password" and click OK.

CONFIGURING MICROSOFT INTERNET EXPLORER

Like Netscape, Microsoft Internet Explorer comes fully configured for normal Web browsing. All you have to do is type in the information about your e-mail and newsgroup accounts. At this writing, Internet Explorer, unlike Netscape, still does not contain an actual mail and news reader within the Web browser itself. Instead, Microsoft bundles two applications called Internet Mail and Internet News with the "full" version of Microsoft Internet Explorer. To configure Internet Mail, simply run the program and answer the questions asked by the Mail Configuration Wizard. Internet News is also configured via a Configuration Wizard the first time it is run.

GOING ONLINE

Now you're ready to go online. To connect to the Net just double-click the icon for your connection in **Dial-Up Networking Folder** (you named it earlier), and click **Connect**. (Note: If you checked "bring up a terminal window after dialing," you'll need to enter your username and password at the prompts. If you are asked to choose a service type, **PPP** is the most common type of connection. Then, click the **Continue** button at the bottom of the window.)

Here's a checklist for the information you'll need to get from your ISP:

ISP's Domain Name:_____

ISP's Local Dial-in Number:_____

ISP's Domain Name Server (DNS):_____

ISP's News (NNTP) Server:_____

ISP's POP Mail Server:_____

ISP's SMTP Server:_____

Your user ID or username:_____

Your password:_____

Your e-mail address:_____

ISP's home page:_____

**ISP's telephone number
(voice line for tech support):**_____

ISP's e-mail address:_____

APPENDIX F

USING INTERNET SOFTWARE WITH MACINTOSH

Configuring a Macintosh for the Internet is very easy (compared to Windows 3.11), but then again, isn't everything about the Mac easier than Windows 3.11? Here's how to do it:

First, determine which version of the Mac's operating system your computer uses. From the Finder (or main screen) pull down the **Apple** menu and select **About this Macintosh** (or **About this Computer** if it's a clone) and look to see which operating system is installed. It should say System 7 point something. Ideally, it will say System 7.6 (or higher). Anything earlier (such as System 6 point something), and you really should upgrade to System 7. Although it is possible to run some Web browsers, such as Netscape, under System 6.0.5, it isn't easy to do. If you are running an early version of System 7 like 7.0.1, you really should upgrade to the latest system your particular Mac can handle—to OS 7.6 (or higher) if possible.

The reason for the upgrading to a higher operating system is simple. To connect a Mac to the Internet, you'll need two key programs (1) a TCP/IP transport layer (such as Apple Open Transport or MacTCP) and (2) a dialer program (such as PPP or MacPPP). Of course, you'll also need a Web browser package, such as Netscape or Internet Explorer. If your Mac uses System 7.6 or higher, then it already comes with Apple Open Transport (the TCP/IP layer) and PPP. In that case, all you'll need is a browser like Netscape or Internet Explorer.

If your Mac uses System 7.5, it should have MacTCP as part of the operating system. This handles the TCP/IP part, so all you would need would be a dialer like MacPPP or PPP. Your ISP will probably furnish you with one of these. If not, you could buy Apple's Internet Connection Kit which includes a PPP program as well as Netscape Navigator and a host of other Internet programs. (Actually, the version of OS 7.5 you really want is 7.5.3 Update 2.0 because it has Open Transport 1.1 which works better than earlier versions.)

If you are using an operating system between 7.0.1 and 7.5.3 Update 2.0, you will need both MacTCP and either MacPPP or PPP. Your ISP may be able to furnish you with these, or you can upgrade to OS 7.6 which includes both. Another alternative would be to get a browser that has both TCP/IP and PPP software included. One is Netscape Navigator's Personal Edition for Macintosh.

If you are using an online service such as America Online or CompuServe, you probably won't need anything extra as long as you get the version of the online service's software specifically designed for the operating system you use. For example, a particular version of AOL's software may require Mac OS 7.5 or higher. In that case, it won't work with earlier versions like 7.0.1.

CONFIGURING TCP/IP

To configure TCP/IP, you'll need to get some information from your ISP (you won't need to do this if you are using an online service like AOL). So you'll already have to have an account established. When you set up an account with an ISP, they will have certain information you'll need to know to configure the software for your Mac. Use the form at the end of this section to record that information. The list below covers the most common questions and most typical answers, however, your ISP may need different information or have different answers.

Connect Via: **PPP** (usually PPP unless you are on a network at the office)

Configure: **Using PPP Server** (unless your ISP tells you otherwise)

IP Address: **will be supplied by server** (unless your ISP tells you otherwise)

Subnet mask: **will be supplied by server** (unless your ISP tells you otherwise)

Router address: **will be supplied by server** (unless your ISP tells you otherwise)

Name Server Address: A group of four numbers, separated by periods, such as 123.45.678.90. Your ISP will have to give you this information.

Domain Name: Usually your ISP's domain name such as myisp.net. Again, only your ISP knows for sure.

CONFIGURING PPP

This one is easy. All you have to do is add your username (which is the same as your e-mail name) and your password. Since these have to match with the username and password stored on your ISP's server, you will have to set these up with your ISP.

▼ *Appendix F*

This number is *all* of the numbers you'll have to dial in order to call your ISP's local access line. You may have to dial 9 to get an outside line or you may want to add a *70 (or some other code) to disable call waiting. Again, your ISP will have to tell you the actual phone number for its access line.

When configuring PPP, select **Options** and **Connections** and make sure "Connect automatically when starting TCP/IP applications" is checked. This means that PPP will start connecting you to the Internet whenever you open your browser.

CONFIGURING YOUR BROWSER

Once you've installed and configured a TCP/IP program and a PPP program, you must now install and configure a browser like Netscape Navigator or Internet Explorer. This is not very hard. All you usually need to do is to tell your browser's e-mail system what your e-mail address is, the name of your ISP's e-mail server, and the name of your ISP's news server. Again, your ISP will need to give you this information.

CONFIGURING NETSCAPE

To configure Netscape, start it and pull down the **Options** window and select **Mail and News**. Now, select **Servers** and type in the names of your ISP's mail and newsservers. Then select **Identity** and type in your name, your e-mail address and your return e-mail address (which is probably the same as your e-mail address). While these are the minimum configuration steps, there are many other features you can customize to your liking (such as changing the default home page—see page 77).

CONFIGURING EXPLORER

Internet Explorer comes with a separate configuration program called Internet Config which makes configuring Explorer very easy. First, start Explorer and click on **Preferences**. At the bottom left-hand corner of the **Preferences** window, you'll see a button marked **Open Internet Config**. Click on that and you'll get more buttons. The ones you need to use are **Personal**, **Email** and **News**. Under **Personal**, you'll need to type in your real name. Under **Email**, type in your e-mail address, e-mail account (probably the same as your e-mail address), password and the SMTP host. Under **News**, type in the NNTP host. The username and password will probably be the same as your e-mail name and password. While these are the minimum configuration steps, there are many other features you can customize to your liking in both **Internet Config** and the **Preferences** window (such as changing the default home page—see page 77).

Here's a checklist for the information you'll need to get from your ISP:

ISP's Domain Name:_____

ISP's Local Dial-in Number:_____

ISP's Domain Name Server (DNS):_____

ISP's News (NNTP) Server:_____

ISP's POP Mail Server:_____

ISP's SMTP Server:_____

Your user ID or username:_____

Your password:_____

Your e-mail address:_____

These would be helpful, too...

ISP's home page:_____

ISP's telephone number
(voice line for tech support):_____

ISP's e-mail address:_____

HELP!

The Mac is so simple to use, you probably won't have any problems, but if you do, give your ISP's Customer Support staff a call.

APPENDIX G

WINDOWS 98/ INTERNET EXPLORER 4.0

As this book went to press, we were not certain exactly when Microsoft would finally release Windows 98. Although we've played around with the Beta "test" versions of Windows 98, we have not seen it in its final form. Undoubtedly, it will be an improvement over Windows 95. For Internet users, the most noticeable change (we think) will be caused by the release of Internet Explorer 4.0, the Internet browser designed specifically for use by Windows 98 (but which will also run under Windows 95).

Internet Explorer 4.0 has all the functionality of Internet Explorer 3.01 and adds many new features which help make the Web much more accessible from your Windows 95/98 desktop.

Perhaps the most important of these new features is Internet Explorer's shell integration. Instead of allowing Internet Explorer to remain a separate, stand-alone application (much like Netscape is today), Microsoft decided to build Explorer right into its Windows Operating System. Now, the same program that allows you to view folders and files on your hard drive (Windows Explorer) is also a built-in Web browser (Internet Explorer). This is an incredible advantage for Explorer because users no longer have to load a separate application to view Web pages. Instead, simply type a URL into ANY open Explorer window (such as "My Computer"), and immediately the window will transform into a Web browser and will go online and load the desired Web page.

Not only does this integration make Explorer much easier to use, but it also allows experienced users to view any folder on the hard drive as if it was a Web page. Simply by choosing "View, Customize this Folder…" users can add an HTML background or choose a custom background
picture for any folder. While this may be confusing at first, it simply means that there will be little visual difference between what you see on your computer and what you see on the Internet.

Several other improvements have also been made in the Internet Explorer browser itself. The first of these involves the addition of "Explorer Bars." Explorer Bars are located on the left of the browser window. Currently, there are four types of Explorer Bars. The Search Bar makes it easier to use a search engine and browse through the results. The History Bar helps you track where you've been today, last week and even last month. The Favorites Bar helps you organize your favorite sites, and the Channels Bar lets you choose which "Channel" you'd like to visit.

The Channels Bar is one of the most talked about new features of Internet Explorer because it is based on the new "push" technology. Most of us "pull" information from the Internet, meaning that we have to know what's there and how to get it. With "push" technology, you tell a service (like PointCast) what information you are interested in, and as that information gets placed on the Web, it is "pushed" on to your computer. With the Channel Bar, you get automatic access to special Web sites that contain information of particular interest to you. For example, an Internet Service Provider or company might put together a package of stories, news articles and other information about a particular subject, much like a cable channel might cover news, weather or sports. Currently, several channels are pre-installed with Internet Explorer 4.0, but you will have the ability to add more channels as they become available. A few of the channels available today are: PointCast , The Microsoft Network, MSNBC News, Disney, C-Net, The Wall Street Journal, Fortune, The Quicken Financial Network, Hollywood Online, AudioNet, National Geographic, The New York Times, Live Wired, ZD Net, ESPN Sportszone, CBS SportsLine, and Warner Brothers Online.

In addition to the integrated Web browser, Internet Explorer 4.0 comes with a number of small, but incredibly useful add-ons to your Windows 95 system. One of these is Microsoft FrontPad, a smaller, free version of the Microsoft FrontPage HTML editor. Considering its cost, FrontPad is a very good, very basic WYSIWYG (What You See Is What You Get) HTML editor, which will allow anyone to create basic Web pages on their own. Like FrontPad, Microsoft Outlook Express in Internet Explorer 4.0 is a smaller, free version of the Office '97 program Microsoft Outlook. Outlook Express is both an e-mail and newsreader, which is designed to compete with the Mail and News Client built into Netscape Communicator. When you start it up the first time, Outlook Express runs its own Set-up Wizard which gathers all the information it needs to download your e-mail and access your ISP's primary news server. Life on the Net just gets easier and easier!

Internet Explorer 4.0 was just released as this book when to press. As Microsoft has pledged, this and all future versions of Internet Explorer will be available for free (as an add-on to Windows) from the Internet Explorer 4.0

Home Page at **http://www.microsoft.com/ie**. However, it will take approximately five hours using a 28.8 bps modem to download the "free" version of Internet Explorer 4.0—you may want to get it on CD-ROM. To get Internet Explorer on CD-ROM, go to **http://www.microsoft.com/ie** (Microsoft's Web site) and follow the link labeled "Order Internet Explorer 4.0 on CD." You will be asked for a credit card number to cover the shipping fee (about $4.95 at this writing).

INDEX

Accessing
    e-mail, 37
    gopher sites, 79
    a Web site, 78-79
    the Internet from rural areas, B-1
Address. See URL.
Address (Web and Gopher), 78-81
Address book (e-mail), 43
Addresses (e-mail), 28, 37-38
Agricultural and farming links, 123
Agricultural companies online, 129
Agricultural magazines online, 142
Agricultural newsgroups, 146
Agricultural organizations online, 143
Agricultural resources, 123
    companies online, 129
    entomology, 173
    farmers online, 174
    forestry, 178
    land grant universities, 179
    livestock resources, 200
    mailing lists, 215
    magazines online, 142
    management and marketing, 215
    market and price information sites, 216
    newsgroups online, 146
    organizations online, 143
    pesticides, 219
    precision farming, 220
    software online, 147, 241
    soil and water, 223
    state departments of agriculture, 225
    turf management, 228
    wildlife, 230
Agricultural software online, 147
Agricultural weather, 275

Alternative agriculture sites, 148
America Online, 6, 26-27, A-1
Anonymous FTP, 86-87
Apples, 154
Arts and science, 230
Attachments (e-mail), 48
Avoiding busy times, B-24
Basic equipment you'll need, 13-29
BBSs. See Bulletin Board Systems.
Beef cattle, 203
Berries, 156
Billing intervals, B-10
Bits per second (bps), 21-22
Bookmarks, 80, B-20
Browsers. See Web browsers.
Browsing
    for FTP sites, 86-87
    for newsgroups, 55-56, 62-64
    the Web, 11, 81-84
Bulletin Board Systems (BBSs), B-12-14
    advantages & disadvantages, B-14
    finding, B-12-13
    how they work, B-13
    logging onto, B-13
    services provided, B-12
    software, B-13
Business and finance, 232
Buying a computer
    new, C-1-3
    sources, C-2-3
    used, C-3-4
Buying a modem, 21-24
Buying software, 14-19
Cache memory, 84
Caching, 83-84
Call waiting, 26
Calling plans, 25-26
Calling zone
    expanded, B-6-7
    LATAs, B-9

Campus MCI (ISP), A-2
Canola, 157
Cattle. See beef cattle or dairy.
Changing
    your e-mail address, 38
    your home page, 84-85
Chat rooms
    danger in, 69-70
    defined, 10
    finding, 68
    how to participate in, 115
    Internet Relay Chat (IRC), 67-68
    software, 67-68
Check on Send, B-18
Choosing an ISP, 27
Citrus fruits, 158
Coffee, 160
Commercial services. See online services.
Companies. See agricultural companies.
Compression. See data compression.
CompuServe, A-3
Computer magazines, 238
Computer resources, 240
    electronic magazines, 244
    hardware companies, 246
    IBM compatible, 248
    Internet Service Providers, 248
    Macintosh, 249
    magazines, 238
    retailers, 240
    software, 241
Computer retailers, 240
Computer software, 241
Computers
    buying new or used, C-1-4
    buying used, C-3-4
    configuring to access Internet, 16-19
    protecting against surges, 19
    recommended, C-2-3
    viruses, 33
    where to buy, C-2-3

Concentric Network (ISP), A-3
Connecting to the Internet
    basic equipment, 13-29
Connection, making a low-cost Internet, A-1
Connection speed (for modem), 24
Connection Wizard, 12
Corn, 160
Cost-cutting strategies, B-14-23
Costs
    billing intervals, B-10
    calling zone, B-6
    comparison (local vs. in-state vs. interstate), B-5
    comparison (urban vs. rural), B-1
    cost-cutting results (table), B-16
    expanded calling zone, B-6-7
    FX vs. toll-free, B-7
    in-state vs. interstate long distance, B-9
    long-distance charges, B-10
    making a low-cost connection, B-5
Cotton, 161
Creating an address book, 43
Crop resources
    apples, 154
    berries, 155
    canola, 156
    citrus fruits, 157
    coffee, 158
    corn, 160
    cotton, 161
    fruits, 164
    general links, 149
    hay and pasture, 162
    industrial crops, 164
    peanuts, 165
    rice, 165
    small grains, 168
    sorghum, 169
    soybeans, 170

tobacco, 171
tubers, 171
vegetables, 171
CSLIP (Compressed Serial Line Internet Protocol), 14
Dairy, 206
Data compression, B-21
Dialer program, 15, 17-19
Domain name, 37
DOS-based PC, 17
Downloading files (FTP), 86-87
Economic development, 250
Education and reference, 254
Electrical surges, 19
Electronic magazines (E-Zines), 244
E-mail
    addresses, 28, 37-38
    and mailing lists, 35
    attachments, 47
    changing your address, 38
    check on send, B-18
    creating an address book, 43
    defined, 10, 35
    how it works, 39
    passwords, 38-39
    reading and sending, 40-47
    reading and writing offline, B-15-17
    reducing costs, B-18
    saving, 44
    sending and receiving, 47-48, 39-41
    servers, 39
    signature file, 43, 46
    software, 36-37
    uses for farmers, 1
Emoticons, 66
Encryption systems, 32-33, 85
Entomology, 173
Error correction, 23
Etiquette. See Netiquette.
Expanded calling zone, B-6-7

External modems, 24
E-zines. See electronic magazines.
FAQs (Frequently Asked Questions), 59
Farm Credit Services, 3-4
Farm Journal Today
    AgFinder™, 113
    Chat Rooms, 112-113
    Discussion Groups, 112
    Monday Night Campfire, 113
    Web site, 2, 111
Farmers online, 174
Favorites (see also Bookmarks), 80, B-20
Fetch (FTP program), 86
File Transfer Protocol. See FTP.
Finding
    an ISP, A-1
    information on the Web, 105
    internet software, 14-19
    local access numbers, B-1
    mailing lists, 51, 215
    newsgroups, 59, 146
    URLs, 105, 123
Flaming, 66
Flat rate service (telephone), 25
Foreign exchange (FX) lines, B-7
Forestry, 178
Forums. See newsgroups.
FreeNets,
    advantages & disadvantages, B-12
    defined, B-11
    services provided, B-11-12
    software, B-12
Frequently Asked Questions. See FAQs.
Fruits, 164
FTP (File Transfer Protocol)
    addresses, 86
    anonymous file transfer, 87
    defined, 86
    how it works, 86
    software, 86

Full-service Internet accounts, 6-8, 26-27
FX. See foreign exchange.
Garden. See home & garden sites.
Glasgow Electric Plant Board, B-4
Goats, 207
Gopher (gopher sites), 79
Government, 255
Grains. See small grains.
Graphical user interface, 11
Graphics, turning off, 83
Handshaking, 30
Hardware companies, 246
Hay and pasture, 162
Health, 258
History, 260
Hogs. See swine.
Home and garden, 261
Home pages, 2, 82
    changing, 84-85
        Farm Journal Today, 2, 111
    on the Web, 84-85
Horses, 208
Host, 6
Hot agricultural & farming links, 123
Hot links. See hyperlinks.
How farmers use the Internet, 2-4
HTML (HyperText Markup Language), 73, 77
HTTP (HyperText Transfer Protocol), 73
Hyperlinks (hypertext links), 77, 82
IBM compatible computer resources, 248
Image maps, 83
Industrial crops, 164
Information appliance, 19
InfoSeek, 108
Installing software. See Setting up software.
Internal modems, 23
Internet
    connecting (requirements), 14
    cost-cutting results (table), B-16
    defined, 5-6
    farming resources on, 119
    five essential services, 10
    netiquette, 66
    search tools, 105-110
    security, 31-32
    setting up software, D-1, E-1, F-1, G-1
    software, D-1-5
    software protocols, 14-15
    what you'll need, 13-29
Internet Config, 57
Internet connection alternatives, B-11
Internet Explorer, 16, 32, 44-46, 51, 57, 62, 74, 81-82
Internet Relay Chat (IRC), 67-68
Internet Service Providers (ISPs), 6
    alternative providers, B-11
    BBSs, 12-14
    changing, 37-38
    choosing, 12-13
    cost of service, B-16-17
    defined, 6
    finding, A-1
    finding local, B-3-4
    FreeNets, B-11-12
    lists of, 245, A-1-4
    national ISPs, 6
    online services vs. national ISPs, 6, 26-27
    online services, 6
    setting up software, 14-19
    toll-free access lines, B-8
    types of, 26-29
Internet services
    five essential, 10
    uses for, 10
Interstate long distance, B-8-9
ISDN, 7, 23
Keywords, 64, B-22
Land grant universities, 179

LATA (Local Access Telephone Area), B-9
Law, 263
Leased lines. See foreign exchange.
Lists
    of ISPs, A-1-3
    of search tools, 107-110
Listserv. See Mailing lists.
Livestock resources, 200
    beef cattle, 203
    dairy, 206
    general links, 200
    goats, 207
    horses, 208
    other livestock, 210
    poultry, 211
    sheep, 212
    swine, 213
Local access number
    connection speed, B-7
    defined, B-1-2
    finding, B-2-4
    if one is unavailable, B-1
    search tips, B-7
Local Internet access providers, B-3
Logging onto a BBS, B-13
Long-distance carriers, B-9-10
Long-distance charges
    billing intervals, B-10
    discount plans, B-9
    expanded calling zone, B-6-7
    in-state vs. interstate, B-8-9
    results table, B-16
    vs. toll-free line, B-8
Lurking, 59
Lycos, 108
Macintosh resources, 249
Macros, 33
Magazines. See Agricultural magazines or computer magazines.

Mail order security, 31-32
Mailing lists
    and e-mail, 49
    compared to newsgroups, 58
    defined, 10, 49
    finding, 51
    moderated vs. unmoderated, 50
    on the internet, 215
    posting articles, 50
    quitting, 53
    reading offline, B-18-19
    requirements, 49
    subscribing, 52
Making a low-cost connection, B-5
    BBSs, B-13-14
    billing intervals, B-10
    expanded calling zone, B-6-7
    FreeNets, B-11-12
    FX line, B-7
    interstate long distance, B-8-9
    other steps to reduce costs, B-18-23
    picking a long-distance plan, B-9-10
    results you can expect to achieve, B-16
    toll-free access lines, B-8
    working online vs. offline, B-14-17
Management and marketing sites, 215
Market and price information sites, 216
MCI Internet (ISP), A-3
Memory, cache, 83-84
Microsoft Internet Explorer. See Internet Explorer.
Microsoft Network, 26-27
Mindspring Enterprises Inc. (ISP), A-3
Modems
    buying, 22
    connection speed, 22-24
    data compression, 23
    error correction, 23
    features, 23

handshaking, 30
how they work, 7, 21
internal vs. external, 23-24
recommended, 22
requirements, 21
serial communications port, 24
speed, 21-23
V.32 and V.42 bis, 23
Moderator, 50, 56
National ISPs
and e-mail, 28, 39
compared to online services, 27-28
costs, 13
defined, 6, 26-27
list of, A-1-3
software provided, 27-30
software setup, 30
National online services. See Online services.
Natural resources, 264
Navigating the Web, 73, 78-82
Netcom Internet Services (ISP), A-3
Netiquette, 66
Netscape, 16, 32, 41-43, 51, 57, 61-62, 74, 81
Network, 5
News, 265
News server, 55
Newsgroups
canceling subscriptions, 58
compared to mailing lists, 58
defined, 10
farming, 60
finding, 60
moderated vs. unmoderated, 56
newsreaders, 61-64
posting to, 59
quitting, 58
reading, 62-63
reading offline, B-15-17

reducing costs, B-19
servers, 55
software, 56
subscribing to, 56-58, 61-65
threads, 59
types of, 60
UseNet, 56
Newsreader (newsgroup reader), 36
Noise, 25
Online services,
advantages & disadvantages, 27
America Online, 8, A-1
and e-mail, 47
compared to national IAPs, 6, 12, 27-28
CompuServe, 8, A-2
defined, 6
keywords, 64
official guides, B-22
operating tips, B-21
setting up software, 64
shortcuts, 64-65
software, 64
toll-free access lines, B-5
turning off graphics, 83-84
upgrading software, 75
Web browsers, 75
Online updates to URLs, 2
Operating practices to save money
avoiding busy times, B-24
bookmarks, B-21
cache memory, 84
changing your home page, 84
data compression, B-22
e-mail, B-19
favorites, B-21
for online services, B-21-22
keywords, B-23
mailing lists, B-19
newsgroups, B-19

printing, B-20
reducing operating costs, B-18-23
results you can expect to achieve, B-16
shortcuts, B-23
stopping a download, B-21
time of day, B-24
turning off graphics, B-20
using the Web, B-19
working offline, B-15
PAML list, 215
Passwords, 38-39, 88
Pasture. See hay and pasture.
Participating in online discussions, 67
Peanuts, 165
Pesticides, 219
Phone lines. See telephone lines.
Pigs. See swine.
Posting articles to mailing lists, 50
Posting to newsgroups, 55
Potatoes. See tubers.
Poultry, 211
PPP (Point-to-Point Protocol), 14-15
Precision farming, 220
Printing, B-20
Programs for Internet. See software.
Quitting
    mailing lists, 53
    newsgroups, 62
Reading
    e-mail, 40-47
    e-mail offline, B-15-17
    newsgroups, B-18
    newsgroups offline, B-18
Reducing costs, B-18-23
Replying to e-mail, 45
Requirements
    computer, 13-14
    for connecting to Internet, 12
    hardware, 14, C-1-3

ISPs, 26-28
modem, 21-24
telephone, 24-25
Retailers. See computer retailers.
Rice, 165
RJ-11 telephone jack, 24
Rural access to Internet, B-1-23
Rural electric cooperatives, B-4
Satellite data services (vs. the Internet), 98
Saving e-mail, 44
Saving URLs. See storing URLs.
Search engines,
    defined, 105
    how they work, 105
    InfoSeek, 108
    lists, 108-109, 268-270
    Lycos, 108
    syntax, 105-107
    using, 107
    Yahoo, 108
Search tips for finding local access, B-1-4
Security on the Internet, 31-33
Sending & receiving e-mail, 39-41, 47-48
Servers, 6
Services, essential Internet, 10
Setting up software, D-1, E-1, F-1, G-1
Sheep, 212
Shortcuts, B-23
Signing up with an online service, 26
SLIP (Serial Line Internet Protocol), 14-15
Small grains, 168
Snail mail, 35
Software
    BBSs, B-13-14
    chat rooms, 67
    communications, D-1
    e-mail, 36
    farm-related, 147
    finding Internet, D-1, E-1, F-1, G-1

newsgroup, 56
provided by ISPs, 14-16
required, 14
setting up, D-1, E-1, F-1, G-1
Web browsers, 74
for Macintosh, F-1-5
for Windows 3.1, D-1-5
for Windows 95, E-1-5
for Windows 98, G-1-3
World Wide Web, 74-75
Soil and water, 223
Sorghum, 169
Soybeans, 170
Sports & recreation, 271
State departments of agriculture, 225
State information, 272
Storing URLs, 80-81
Subscribing
to mailing lists, 52
to newsgroups, 56-58, 61-65
Surge protectors, 19
Swine, 213
Syntax for search tools, 105-107
Tax information, 273
TCP/IP, 12, 14-19
Telecommunications, 274
Telephone companies, 276
Telephone jack, 24
Telephone line
call waiting, 26
calling plans, 25-26
getting a second line, 25
local vs. long distance, B-5-10
noise, 25
poor connection, 25
real world throughput, 23
toll-free vs. long distance, B-5-10
Threads, 56, 59
Tobacco, 171

Toll-free access, B-8
Travel, 276
Trumpet Winsock, 15
Tubers, 171
Turf management, 228
Turning off
call waiting, 26
graphics, 83
TVA Rural Studies, 3
Uniform Resource Locator. See URL.
Unlimited access, 8
University resources, 101-104, 272
Unsubscribing. See quitting.
Updates to URLs, 2
URL (Uniform Resource Locator)
defined, 30, 78
entering, 31, 79-80
finding, 105
storing, 80-81
URLs of agricultural resources. See
index on page 119.
UseNet. See newsgroups.
V.32 & V.42 bis, 23
Vegetables, 171
Viruses, 32-33
Weather
Canadian weather, 291
International weather, 291
National Weather Service offices, 284
Regional weather, 282
USA weather, 278
Web browsers
accessing a Web site, 78
and newsgroups, 57
bookmarks and hotlists, 80, B-20
cache memory, 84
defined, 11
Internet Explorer, 16, 32, 44-46, 51, 57,
62, 74, 81-82

Netscape, 16, 32, 41-43, 51, 57, 61-62, 74, 81
    turning off graphics, 83
Web server, 78
Web sites, 78, 119-291
    Farm Journal Today, 2, 111
    farmers, 174
Web. See World Wide Web.
WebTV, 19-20
Wildlife, 230
Working online vs. offline, B-14
    comparison table, B-17
    connection time, B-18
World Wide Web (WWW)
    addresses, 78-81, 119-291
    and the Internet, 11
    browsing the Web, 11, 81-84
    changing your home page, 84-85
    defined, 10, 73
    farming information on, 119
    finding information, 105
    graphical user interface, 11
    home pages, 84-85
    how it works, 76-77
    safety, 85
    software, 74-76
    URLs (defined), 30, 78
    Web sites (defined), 77
Writing e-mail messages, 42-43, 45-47
Writing e-mail offline, B-14-16
WWW. See World Wide Web.
Yahoo, 108

YOUR FAVORITE WEB ADDRESSES